ESTIMATING PROBABILITIES OF EXTREME FLOODS

METHODS AND RECOMMENDED RESEARCH

Committee on Techniques for Estimating
Probabilities of Extreme Floods

Water Science and Technology Board
Commission on Physical Sciences,
Mathematics, and Resources

National Research Council

NATIONAL ACADEMY PRESS
Washington, D.C. 1988

National Academy Press • 2101 Constitution Avenue, N.W. • Washington, D. C. 20418

NOTICE: The project that is the subject of this report was approved by the Governing Board of the National Research Council, whose members are drawn from the councils of the National Academy of Sciences, National Academy of Engineering, and the Institute of Medicine. The members of the committee responsible for the report were chosen for their special competences and with regard for appropriate balance.

This report has been reviewed by a group other than the authors, according to procedures approved by a Report Review Committee consisting of members of the National Academy of Sciences, the National Academy of Engineering, and the Institute of Medicine.

The National Academy of Sciences is a private, nonprofit self-perpetuating society of distinguished scholars engaged in scientific and engineering research, dedicated to the furtherance of science and technology and to their use for the general welfare. Upon the authority of the charter granted to it by the Congress in 1863, the Academy has a mandate that requires it to advise the federal government on scientific and technical matters. Dr. Frank Press is president of the National Academy of Sciences.

The National Academy of Engineering was established in 1964, under the charter of the National Academy of Sciences, as a parallel organization of outstanding engineers. It is autonomous in its administration and in the selection of its members, sharing with the National Academy of Sciences the responsibility for advising the federal government. The National Academy of Engineering also sponsors engineering programs aimed at meeting national needs, encourages education and research, and recognizes the superior achievements of engineers. Dr. Robert M. White is president of the National Academy of Engineering.

The Institute of Medicine was established in 1970 by the National Academy of Sciences to secure the services of eminent members of appropriate professions in the examination of policy matters pertaining to the health of the public. The Institute acts under the responsibility given to the National Academy of Sciences by its congressional charter to be an adviser to the federal government and, upon its own initiative, to identify issues of medical care, research, and education. Dr. Samuel O. Thier is president of the Institute of Medicine.

The National Research Council was organized by the National Academy of Sciences in 1916 to associate the broad community of science and technology with the Academy's purposes of furthering knowledge and advising the federal government. Functioning in accordance with general policies determined by the Academy, the Council has become the principal operating agency of both the National Academy of Sciences and the National Academy of Engineering in providing services to the government, the public, and the scientific and engineering communities. The Council is administered jointly by both Academies and the Institute of Medicine. Dr. Frank Press and Dr. Robert M. White are chairman and vice chairman, respectively, of the National Research Council.

Support for this project was provided by the U.S. Nuclear Regulatory Commission under Contract Number NRC-03-85-064.

Library of Congress Cataloging-in-Publication Data

National Research Council (U.S.). Committee on
    Techniques for Estimating Probabilities of Extreme Floods
    Estimating probabilities of extreme floods.

    Bibliography: p.
    Includes index.
    1. Flood forecasting. I. Title.
GB1399.2.N37 1988   551.48'9   87-34839
ISBN 0-309-03791-3

Copyright © 1988 by the National Academy of Sciences

No part of this book may be reproduced by any mechanical, photographic, or electronic process, or in the form of a phonographic recording, nor may it be stored in a retrieval system, transmitted, or otherwise copied for public or private use without written permission from the publisher, except for the purposes of offical use by the U.S. government.

Printed in the United States of America

COMMITTEE ON TECHNIQUES FOR ESTIMATING PROBABILITIES OF EXTREME FLOODS

JARED L. COHON, The Johns Hopkins University, *Chairman*
VICTOR R. BAKER, University of Arizona
DUANE C. BOES, Colorado State University
C. ALLIN CORNELL, Stanford University
NORMAN H. CRAWFORD, Hydrocomp, Mountain View, California
MICHAEL D. HUDLOW, National Weather Service
WILLIAM KIRBY, U.S. Geological Survey
DONALD W. NEWTON, Tennessee Valley Authority
KENNETH W. POTTER, University of Wisconsin-Madison
JAMES R. WALLIS, IBM Watson Research Center, Yorktown Heights, New York
SIDNEY J. YAKOWITZ, University of Arizona

Ex-Officio (WSTB Members)

STEPHEN J. BURGES, University of Washington
LEO M. EISEL, Wright Water Engineers, Denver, Colorado

National Research Council Staff

STEPHEN D. PARKER, *Project Manager*, Water Science and Technology Board
RENEE A. HAWKINS, *Project Secretary*, Water Science and Technology Board

U.S. Nuclear Regulatory Commission Project Officer

DONALD L. CHERY, JR.

Other Resource Persons

ARLEN D. FELDMAN, U.S. Army Corps of Engineers
WILLIAM L. LANE, U.S. Bureau of Reclamation

WATER SCIENCE AND TECHNOLOGY BOARD

JOHN J. BOLAND, The Johns Hopkins University, *Chairman*
STEPHEN J. BURGES, University of Washington
RICHARD A. CONWAY, Union Carbide Corporation, South Charleston, West Virginia
JAMES M. DAVIDSON, University of Florida
HARRY L. HAMILTON, JR., State University of New York at Albany
JAMES HEANEY, University of Florida
R. KEITH HIGGINSON, Idaho Department of Water Resources, Boise
MICHAEL KAVANAUGH, James M. Montgomery Consulting Engineers, Oakland, California
LESTER B. LAVE, Carnegie-Mellon University, Pittsburgh, Pennsylvania
LUNA B. LEOPOLD, University of California-Berkeley
G. RICHARD MARZOLF, Kansas State University
JAMES W. MERCER, GeoTrans, Herndon, Virginia
GORDON ROBECK, Consultant, Laguna Hills, California
PATRICIA ROSENFIELD, The Carnegie Corporation of New York
EDITH BROWN WEISS, Georgetown University Law Center, Washington, D.C.

Staff

STEPHEN D. PARKER, Director
SHEILA D. DAVID, Staff Officer
PATRICK W. HOLDEN, Staff Officer (through September 1987)
WENDY L. MELGIN, Staff Officer
CAROLE B. CARSTATER, Staff Assistant
JEANNE AQUILINO, Administrative Assistant
RENEE A. HAWKINS, Senior Secretary
ANITA A. HALL, Senior Secretary

COMMISSION ON PHYSICAL SCIENCES, MATHEMATICS, AND RESOURCES

NORMAN HACKERMAN, Robert A. Welch Foundation, *Chairman*
GEORGE R. CARRIER, Harvard University
DEAN E. EASTMAN, IBM T.J. Watson Research Center
MARYE ANNE FOX, University of Texas
GERHART FRIEDLANDER, Brookhaven National Laboratory
LAWRENCE W. FUNKHOUSER, Chevron Corporation (retired)
PHILLIP A. GRIFFITHS, Duke University
J. ROSS MACDONALD, The University of North Carolina, Chapel Hill
CHARLES J. MANKIN, Oklahoma Geological Survey
PERRY L. MCCARTY, Stanford University
JACK E. OLIVER, Cornell University
JEREMIAH P. OSTRIKER, Princeton University Observatory
WILLIAM D. PHILLIPS, Mallinckrodt, Inc.
DENIS J. PRAGER, MacArthur Foundation
DAVID M. RAUP, University of Chicago
RICHARD J. REED, University of Washington
ROBERT E. SIEVERS, University of Colorado
LARRY L. SMARR, National Center for Supercomputing Applications
EDWARD C. STONE, JR., California Institute of Technology
KARL L. TUREKIAN, Yale University
GEORGE W. WETHERILL, Carnegie Institution of Washington
IRVING WLADAWSKY-BERGER, IBM Corporation

RAPHAEL G. KASPER, *Executive Director*
LAWRENCE E. MCCRAY, *Associate Executive Director*

Preface

Other than the familiar death and taxes, few events, if any, can be predicted with certainty. Virtually all of our decisions and actions are taken in the face of uncertainty, from the daily choice of a commuting route to the design of a spillway on a dam or regulatory approval of new drugs. In each of these cases, making a decision requires weighing the consequences of alternative actions and the likelihood of each consequence occurring. When the consequences are relatively unimportant, as in the daily commuting trip, we are usually content to make decisions quickly and informally. An extra 10 minutes in traffic is not a very stiff penalty, though I remember situations when it was hard to summon any equanimity. However, when the consequences are major, as with most public actions, decisions often can be improved by the formality and structure of risk analysis.

Risk analysis is a generic term for methods that support decisionmaking by quantifying consequences and their probabilities of occurrence. The various methods in different settings are called probabilistic risk assessment, risk assessment and management, or, simply, risk analysis. Whether one is talking about nuclear power plants or environmental regulation, the underlying problems are the same: identification of consequences, estimation of probabilities, and the combination and consideration of results prior to decisionmaking.

This report concerns rare floods and the estimation of their probabilities of occurrence. Our interest was in floods with a probability of occurrence of much less than once in 100 years, say 10^{-3} to 10^{-7} chance per year. Our focus was entirely on the estimation of probabilities. We did not perform risk analyses, nor did we review the methods or policy of risk-based decisionmaking. Other studies, especially that of the Committee on Safety Criteria for Dams (National Research Council, 1985), have considered these issues. Indeed, various earlier reports provided excellent motivation for our study and served as our point of departure.

Estimating the probabilities of extreme floods is an important and challenging problem. It is important because the stakes are high: very large floods kill people and destroy property, and the cost incurred in attempting to avoid these damages can be great. The probabilities that such floods will occur during the life of a particular project are a crucial part of the analytical input for making decisions about that project.

Assigning probabilities to extreme floods is challenging because we cannot know for certain the probability that a given flood will occur or even the probability distribution that flood peaks follow. Basically our problem is to assign a probability to an event that may occur, say, once in a million years, using data for, perhaps, the last hundred years. It is a formidable task that one is tempted not to attempt. Yet, decisions must be made, and probabilities should be associated with floods to provide a sound basis for those decisions.

How should these probabilities be estimated? Can these probabilities be estimated with reasonable accuracy? What research should be performed to improve existing methods and develop new ones for probability estimation? These were the three questions that the committee was formed to address.

This report provides answers to those three questions, but we are cautious in our recommendations. We were guided by a basic tenet: it's usually better to have more information, if you can collect the data and make sense of the results. Thus, we considered carefully all the major methods for flood probability estimation, evaluating them for their analytical cogency, physical soundness, and data requirements. Recognizing that no single method could solve the problem, we did not prepare a cookbook of recipes for flood estimation. Instead, we did something more valuable by laying out a coherent framework for flood probability estimation that provides guidance for current practice, a basis for critical evaluation of individual techniques, and directions for future research.

The history of research and practice in flood probability estimation, like other fields in which observation of the object of analysis is difficult or impossible, has been marked by sharp disagreements and well-defined schools of thought. All camps were represented on the committee, and I am pleased to report that they worked together in remarkable harmony. The collegiality and productivity of the group were wonderful, and I thank the committee members and all the others involved (see page iii) for their participation and good work.

JARED L. COHON, *Chairman*
Committee on Techniques for Estimating
Probabilities of Extreme Floods

Contents

1. BACKGROUND AND OVERVIEW 1
2. IMPROVING THE THEORETICAL BASIS 4
   Approaches to Estimation, 4
   Three Principles for Improving Estimation, 6
   Recommended Statistical Techniques, 8
   Recommended Runoff Modeling Techniques, 9
   Uncertainty Analysis, 11
3. FLOOD-BASED STATISTICAL TECHNIQUES 12
   Introduction, 12
   Problem Delineation–Background–Modeling–Notation, 13
   Single Site Estimation, 26
   Multisite/Regional Analysis, 34
   Assessment, 53
   A Strategy for Estimating $\xi_i(q)$, 53
   Conclusion, 54
4. RUNOFF MODELING METHODS 55
   Introduction, 55
   Runoff Models, 57
   Development of Meteorologic Inputs, 58
   Estimation of Probabilities, 73
   Analysis of Uncertainty, 74
   Conclusions, 77

5. DATA CHARACTERISTICS AND AVAILABILITY.........80
   Introduction, 80
   Streamflow Data, 84
   Rainfall Data, 91
   Paleoflood Hydrology, 105
   Additional Characteristics of Flood Data, 111
6. DEVELOPMENTAL ISSUES AND RESEARCH NEEDS..118
   Flood Statistical Analysis, 118
   Runoff Modeling, 119
   Data Considerations, 120
   A Comprehensive Statistical Model, 122

REFERENCES..123
BIOGRAPHICAL SKETCHES OF COMMITTEE
MEMBERS..133
INDEX...137

TABLES

1. Suite of Methods for Estimating Large Floods 5
2. Summary Description of Five Floodsets 41
3. Description of Floodset 1 42

FIGURES

1. Cumulative distributions of $\hat{\mu}$, $\hat{\sigma}$, and $\hat{\gamma}$, for a three-parameter log normal distribution with $\mu = 0$, $\sigma = 1$, and $\gamma = 15$ for $n = 30$ (after Wallis et al., 1974).......................... 24
2. Skew-kurtosis relationship for annual maxima of California point rainfall data estimated by the WAK/R algorithm and with no small sample bias corrections attempted (after Wallis, 1982)... 25
3. At-site GEV quantile estimates (based upon unbiased PWM's) for site 21 of Floodset 2, showing the median as well as the upper and lower quartile and decile values......... 45
4. GEV-1 quantile estimates (based upon unbiased PWM's) for site 21 of Floodset 2, showing the median as well as the upper and lower quartile and decile values. 45

5. GEV-2 quantile estimates (based upon unbiased PWM's) for site 21 of Floodset 2, showing the median as well as the upper and lower quartile and decile values. 46
6. WAK/R quantile estimates (based upon unbiased PWM's) for site 21 of Floodset 2, showing the median as well as the upper and lower quartile and decile values 46
7. GEV-1 quantile estimates (based upon unbiased PWM's) for site 21 of Floodset 1, showing the median as well as the upper and lower quartile and decile values...... 47
8. GEV-2 quantile estimates (based upon unbiased PWM's) for site 21 of Floodset 1, showing the median as well as the upper and lower quartile and decile values...... 47
9. GEV-1 quantile estimates (based upon unbiased PWM's) for site 1 of Floodset 1, showing the median as well as the upper and lower quartile and decile values...... 48
10. GEV-1 quantile estimates (based upon unbiased PWM's) for site 41 of Floodset 1, showing the median as well as the upper and lower quartile and decile values...... 48
11. GEV-2 quantile estimates (based upon unbiased PWM's) for site 1 of Floodset 1, showing the median as well as the upper and lower quartile and decile values...... 49
12. GEV-2 quantile estimates (based upon unbiased PWM's) for site 41 of Floodset 1, showing the median as well as the upper and lower quartile and decile values...... 49
13. GEV-1 quantile estimates (based upon unbiased PWM's) for site 21 of Floodset 5, showing the median as well as the upper and lower quartile and decile values...... 50
14. GEV-1 growth curve estimates (based upon unbiased PWM's), for Floodset 1, showing the median as well as the upper and lower decile values. 51
15. GEV-1 growth curve estimates (based upon unbiased PWM's), for Floodset 3, showing the median as well as the upper and lower decile values. 52
16. GEV-1 quantile estimates (based upon unbiased PWM's) for site 21 of Floodset 4, showing the median as well as the upper and lower quartile and decile values. 52
17. Distribution of major precipitation events for which official (U.S. Army Corps of Engineers) depth–area–duration data have been published. 66
18. Distribution of major precipitation events for which limited information is available. 68

19. Example of completed United States storm-study results—storm location map and maximum depth–area–duration data for the storm of July 9-13, 1951, centered near Council Grove, Kans. 98
20. Example of completed United States storm-study results—total-storm isohyetal map and mass curves for principal rainfall centers for the storm of July 9-13, 1951, centered near Council Grove, Kans. 95

ESTIMATING PROBABILITIES OF
EXTREME FLOODS

1

Background and Overview

At the request of the U.S. Nuclear Regulatory Commission, the National Research Council, through its Water Science and Technology Board, in 1985 initiated a study of techniques for estimating probabilities of extreme floods. An enabling agreement provided that the National Research Council would establish a study committee to (1) review and critique various approaches to estimation of extreme flood probabilities, (2) assess and identify a preferred approach to flood estimation to be further developed (or used now, if possible), and (3) identify research that can be expected to improve our ability to estimate flood magnitudes and probabilities using the recommended approach and other approaches. The scientific methodology needed for estimating the probability of rare floods was the essence of the assignment: it was not to address policy matters, flood damage assessment methods, or flood risks at specific sites.

The Water Science and Technology Board initiated this study in November 1985 with the appointment of the Committee on Techniques for Estimating Probabilities of Extreme Floods. The committee included experts with backgrounds in hydrology, meteorology, applied probability, and statistics. During the study period, the committee met several times (January 20–21, March 11–12, June 2–3, October 20–21, 1986, and March 5–6, and June 2, 1987) for discussions and debates on substantive issues, report writing sessions, and reviews. In the period following the committee's last meeting and until report publication, considerable effort was spent in improving the quality of this report to the satisfaction of the committee and the National Research Council's Report Review Committee because it was hoped at the outset that the report would serve hydrologists as a valuable reference for years to come.

The request from the U.S. Nuclear Regulatory Commission that the National Research Council undertake this study is but one more indication of the practical need to be able to assign probabilities to the occurrence of rare floods. Risk assessment and economic evaluation methods have advanced to the point where improved estimates of the probabilities of extreme floods are being called for in a variety of planning and design situations. For example, the safety features of dams and nuclear power plants are designed on the basis of floods up to and including the probable maximum flood (PMF)—the flood that can be expected from the most severe combination of critical meteorologic and hydrologic conditions that are reasonably possible in the region. Previous studies of dam safety by the National Research Council (1983b, 1985) have indicated the need for methods for estimating probabilities and uncertainty bounds for extreme floods ranging from return periods of several thousand years up to the PMF.

A work group of the Hydrology Subcommittee of the Interagency Advisory Committee on Water Data investigated the feasibility of assigning probabilities and uncertainty bounds to floods of the order of magnitude of the PMF (Hydrology Subcommittee, 1986). The work group examined the hydrologic and engineering literature on various methods of flood probability estimation, including joint-probability, regional-data, and paleohydrologic methods. It identified and discussed several obstacles to the use of available hydrometeorological data sets for defining the extreme tails of flood probability distributions and concluded that the state of the art of extreme-flood probability analysis had not developed to the point where a method could be implemented for operational use.

Floods of concern here include those with annual probabilities in the broad range of from 10^{-2} to "near 0." However, in the United States streamflow records of greater than 100 years are meager, and most records are considerably shorter. Consequently, any method for estimating probabilities of floods rarer than about the 100-year flood must include some form of extrapolation, a process that can, at best, introduce errors and, at worst, strain credulity.

It is precisely because extreme floods are rare that it is difficult to quantify their probabilities. At the same time, increasing emphasis is being given to accounting more precisely for the risk of failure of engineered structures. Decisions based on risk analysis have become more common in science, engineering, and public decisionmaking, yet techniques for estimating the probabilities and associated confidence limits of rare floods need to be improved considerably. This point of view is strongly supported by the Water Science and Technology Board in its Annual Reports for both 1983 and 1984, and also by the Board's Committee on Safety Criteria for Dams (National Research Council, 1985). It is hoped that this report will further the science of rare flood hydrology.

In the course of its work, the committee identified and reviewed various available approaches to the estimation of extreme flood magnitudes and probabilities. Such approaches include several statistical techniques based on analysis of systematic streamflow records, techniques based on analysis of joint probabilities of individual events, and techniques based on analysis of precipitation records, use of historical and paleoflood data, and regionalized approaches. The committee's review focused on the theoretical soundness and scientific basis of each approach and concentrated mainly on available literature and data. Neither a thorough and systematic comparative analysis nor data testing were carried out. Data were, however, acquired and analyzed in some cases where required for comparison purposes.

The committee concluded that there are opportunities to improve the practice and science of rare flood hydrology. As a point of departure, the committee observed that a framework that allows a range of floods and their probabilities to be estimated is preferred to an approach that focuses only on a single, large flood that is intended to represent some sort of "upper bound." The framework should be based on statistical and rainfall-runoff modeling methods. No single technique is clearly preferable to all others, and there are some techniques that probably cannot be recommended in most situations. In the short term, the committee identified and recommended an approach that makes use of the best of the existing methods. Chapter 2 summarizes this broad approach to flood probability estimation and includes suggestions that will require limited further research or development to make implementation possible. The many details of the various techniques that are important in understanding their use and the context for any recommendations are discussed in chapters 3 and 4. Data are important for implementing any of the techniques, and data availability is discussed in chapter 5. Chapter 6 presents several topics for further development and research that can improve on the recommended approach to flood-frequency estimation.

2

Improving the Theoretical Basis

Extreme or rare floods, with probabilities in the range of 10^{-3} to 10^{-7} (more or less) chance of occurrence per year, are of continuing interest to the hydrologic and engineering communities for purposes of design and planning. When compared to the very long return periods of interest, the historical record of such events is small; thus opportunities to test or compare estimated flood quantiles with experienced events almost never occur. Nevertheless, the need to design or plan for the occurrence of extreme floods is real. The committee believes that advances in the probabilistic modeling and statistical analysis of extreme events have been made and that these advances can be applied to improve extreme-event hydrologic analyses so that estimates of the probability of extreme hydrologic events will become possible.

APPROACHES TO ESTIMATION

There are many different methods that could be used to estimate magnitudes of extremely rare floods. The three general types of methods examined by the committee are summarized in Table 1. We note that although the first method, the deterministic estimation of "probable maximum floods," is used worldwide for engineering design, it does not provide the probabilities needed in risk-assessment work. Consequently, the committee focused its attention on the other

IMPROVING THE THEORETICAL BASIS

TABLE 1 Suite of Methods for Estimating Large Floods

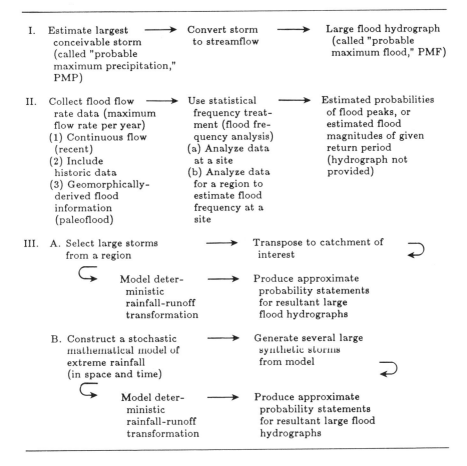

two methods, which provide both flood magnitudes and estimates of associated probabilities.

The committee endorses the concept that the objective of flood studies should be to generate as much information as practicable about the range of flood potential at a site. This is consistent with the idea that choices of the dimensions of engineered structures or planning criteria should be based on an evaluation of the risk and consequences of the decision. Not including the PMF approach, there are two broad categories of approach for estimating ranges of flood potential from available data: those employing statistical analysis of streamflow data, and those employing statistical analysis of meteorological data and some type of model to simulate the physical processes of runoff. The committee did not identify a specific algorithm for computing the probabilities of floods of interest. Rather, this chapter summarizes some principles and two general recommended approaches that should both be considered in flood studies. Flood probability estimates should usually be made employing both approaches because increased insight relevant to decisionmaking will be gained from both types of information. While a problem may remain in relating the presumably different estimates, further research and development should improve our ability to combine results from the two approaches (see chapters 3 through 5).

THREE PRINCIPLES FOR IMPROVING ESTIMATION

Estimating the probabilities of extreme floods will always require extrapolation well beyond the data set, and guidance for doing this is not well defined. Even methods that extend records, such as using paleohydrologic data, do not provide enough information for estimating flood probabilities on the order of 10^{-3} or smaller. Faced with this difficulty, we have identified three principles for improving extreme flood estimation. These principles, applicable to both statistical analysis of streamflow and hydrometeorological modeling, are: (1) "substitution of space for time"; (2) introduction of more "structure" into the models; and (3) focus on extremes or "tails" as opposed to or even to the exclusion of central characteristics. In addition, the interest in extreme floods increases the importance of explicit uncertainty analysis.

The first principle, substitution of space for time, is useful where the dynamic hydrological characteristics of the site of interest are similar to those of a broad region. If differences and interdependences are properly accounted for, this regional information can effectively increase the data base at the site. The magnitude of this increase, the degree of reduction of estimation error, and the extent of additional information about rare events will depend on the number, homogeneity, and degree of statistical independence among data records in the region. In flood frequency analysis (chapter 3), this step is represented by "regionalization," i.e., incorporating in the analysis information from other flood data sets from other locations, streams, and watersheds in the same region. In rainfall-runoff modeling (chapter 4), the substitution is accomplished in different ways by different methods, for example, by probabilistically transposing storms observed in the meteorologically homogeneous region.

The second principle is the introduction of more structure into statistical and simulation models or into the process of regionalization. In the case of regionalized flood-frequency estimation (chapter 3), this structure appears in two ways: first, in the nature of the spatial stochastic dependence assumptions, and second, and more importantly, in the assumption of definite relationships among the regional parameters. As discussed in chapter 4, analogous assumptions may be built into stochastic rainfall models and regionalized intensity–duration–frequency curves (IDFs); in the stochastic storm transposition methods, there are stronger assumptions about the event recurrence model (e.g., Poisson) and the homogeneity in space of the mean arrival rate. These stronger assumptions put added emphasis on the need to include the model uncertainty in the total uncertainty analysis along with the more familiar and more widely reported parameter uncertainty for a given model. However, the information to support these stronger modeling/structure assumptions comes from more than local data. The continuity of nature is often presumed, so that if a certain structure has been verified in other areas, it may be valid in the region of interest; however the assumption should be critically evaluated. The assumptions included in such models make the assessment of model uncertainty difficult to carry out in any statistically rigorous manner.

The third principle, the focus on "tails" or extreme events, is self-evident and is the basis of probable maximum flood practice brought here to the probabilistic context. It is based on the observation that hydrometeorological and watershed processes during extreme events are likely to be quite different from those same processes during more common events. Under extreme conditions, for example, assumptions about watershed physical conditions, such as soil moisture, may become less critical, or stream hydraulics may change. But the focus on tails may also simplify some aspects of the problem. For example, data collection and analysis for more common events can be de-emphasized, and paleodata may exist at the upper flood levels, effectively extending the amount of relevant available data. In chapter 3, this principle manifests itself in the recent trend towards the use of tail-weighted estimation schemes. In chapter 4, we see methods, such as storm transposition, that use *only* the data from extreme storms.

RECOMMENDED STATISTICAL TECHNIQUES

In any flood study, the analyst should make use of all information available, including at-site streamflow and raingage data, regional streamflow data, regional storm data, available historic and paleoflood data, and available probable maximum flood estimates. The flood-frequency assessment should address a range of flood potential at a given site. Initially, a single at-site analysis can be performed. This is a good starting point, and it uses recorded data at the point of interest. However, it must be recognized that data at a single site are too limited to permit more than a rough estimate and then only for relatively common floods.

In the "at-site" category, both parametric and nonparametric analyses may be performed. In the case of parametric analyses, annual peak flow data should be assembled first. The Hydrology Subcommittee (1982) has issued a set of guidelines known as Bulletin 17-B, which is followed by federal agencies for assembling, displaying, and analyzing annual peak flow data using the log-Pearson Type III distribution. A histogram or probability plot can be prepared to perform a crude goodness-of-fit analysis. This might help to eliminate some obviously unsuitable parent distributions while allowing many others. For the allowable parent distributions, estimates of and standard errors for quantiles of interest should be obtained. There are numerous probability density functions. The log-Pearson Type III is one and it is generally used by federal agencies. Another is the generalized extreme value distribution (GEV) that is used in the United Kingdom and elsewhere, and under some conditions can be theoretically supported. Still another is the Wakeby, which possesses substantial flexibility. These would make good starting parent distributions. Partial duration series analysis is part of the "tail" methods and is the heart of Smith's approach (see chapter 3). For nonparametric single-site analyses, we suggest performing various tail analyses including the extrapolation method proposed by Breiman and Stone (see chapter 3).

After these preliminaries, the emphasis should be on increasing the data pool as much as practicable. There are two ways to do this: use of historical and other data, and use of regional analysis. All historical and other data should be considered and used if they are of acceptable quality and relevant to the location of interest. Information obtained by paleohydrologic methods is an example of the type of data that can be obtained; the committee recommends its consideration. It may prove useful to perform a reconnaissance study to determine whether paleoflood information of appropriate quality can be produced at reasonable cost. While sometimes problems exist in making these data systematic for direct incorporation into statistical techniques (see chapter 5), historical and paleoflood data can be useful for comparison purposes.

Geographic regional analysis is another way to extend the data set for estimating the annual peak flood distribution. First, a "largish" data set, where "largish" has not been specifically defined, for the "region" should be assembled, using site descriptors as an aid. The assemblage of the regional data set should provide some of the following desirable characteristics. Some degree of homogeneity is essential. Little serial correlation is expected but, if present, needs to be accounted for. Further, if nonnegligible cross-correlation exists, a few-parameter surface fit may be performed to the correlation function defined over the region and an appropriate joint distribution used. The absence of substantial cross-correlation makes the next step easier. A "structure," which is an assumption about the parameters in the distribution of peaks at the sites in the region, is necessary and its selection is still a major hurdle. An index flood scheme with a GEV parent as described in chapter 3 is a starting point. Research on alternatives is needed. Next, estimates of the quantile function and its standard error are obtained. Simulation and bootstrap methods may be useful for estimating statistical sampling errors. [A useful starting point for those unfamiliar with jackknife and bootstrap procedures is Efron (1982).] Finally, new estimates should be obtained using parent distributions other than GEV but which appear nearly as likely, given the characteristics of the observed data. Such comparisons might help to provide an assessment of the model's robustness.

RECOMMENDED RUNOFF MODELING TECHNIQUES

While discharge (runoff) models have not often been used to generate flood-frequency distributions for ranges far beyond the 10^{-2} annual-probability event, they have been used since their inception to define extreme flood events, such as the probable maximum flood. The committee believes that runoff models, in conjunction with information on magnitude and frequency of extreme meteorological events, have considerable potential as tools for providing estimates of the exceedance probabilities of very rare floods. Runoff models, with given meteorological inputs, provide for simulation of the physical processes that produce streamflow. Such models generate entire hydrographs, often useful in design, and can be applied where streamflow data are limited. Potentially, their greatest advantage is that they permit a separate consideration of the meteorological factors that cause flooding and the watershed response that establishes the runoff characteristics of peak and volume. This in turn permits regionalization of meteorological factors that are more similar over broad regions.

Chapter 4 contains considerable detail about the choice and application of runoff models. While numerous "continuous" or "event" models are available, the committee did not critique them. The committee recommends that the chosen model should simulate the physical processes that will occur during very large events that are outside the range of its calibration.

A critical choice in runoff modeling is the type of meteorological input to be used, i.e., the choice among direct use of historic precipitation data including both station and storm records, stochastically generated storms, or synthetic storms. Of these, the committee believes the synthetic-storm approach is most promising for application and future development. There are two general types of synthetic-storm development: transposition of historic storm information, and regionalization of depth–duration–frequency station data. The committee recommends storm transposition methods, which are described in detail in chapter 4. We note, however, that there are major geographical areas where the historic storm catalog—maintained by the U.S. Army Corps of Engineers—is limited. Storm transposition will not be possible in these regions until they have longer and more complete records of major storms. In areas where the storm catalog is limited, regionalization of station data is recommended as a way of obtaining meteorologic inputs to the chosen event runoff model. The committee recognizes, however, that both storm transposition and other methods of regionalization may be difficult in areas with complicating factors such as orographic effects. Improvements in hydrometeorological methodology may be needed to permit effective use of synthetic storm runoff modeling in such areas.

Because the use of a synthetic-storm approach requires an event runoff model (as opposed to a continuous model), the estimation of antecedent conditions may be critical. Depending upon the region and the watershed, such conditions can include soil moisture, reservoir conditions, and snowpack. Use of a continuous accounting model is recommended in estimating a frequency distribution for soil moisture. The committee did not explicitly address the matter of initial reservoir conditions (or upstream dam failures); judgments must be made based on an evaluation of the expected conditions for the storm being postulated. Though snowpack melt is important to runoff generation in some regions, the committee did not make an intensive study of the methods used to model this phenomenon.

The calibrated runoff model, using meteorologic and antecedent condition inputs, will produce flood hydrographs as outputs. Probabilities can be estimated by integrating the joint probabilities of storm and antecedent condition inputs.

UNCERTAINTY ANALYSIS

A thorough study of low-frequency extreme events requires inclusion of an explicit uncertainty analysis. Sources of uncertainty are many and include sampling uncertainty, measurement errors, and modeling assumptions. Some aspects of an uncertainty analysis, such as determination of the standard error of the estimator of a quantile given the underlying model, are familiar and are discussed in chapters 3 and 4. Other aspects, such as incorporation of model uncertainty, are less familiar. An assessment of relative weights (degrees of belief) on alternative distributions and/or models may be needed. Also, several types or philosophies of uncertainty analysis are possible (Statistical Science, 1987); they include, for example, use of certainty factors and belief functions, use of fuzzy set theory, or use of Bayesian methods. An uncertainty analysis requires that several uncertainties be combined to yield net uncertainty, and the method for combining must be reckoned with. The model-to-model uncertainty may dominate parameter uncertainty in extreme natural event analysis.

Uncertainty pervades many fields and disciplines; currently the calculus of uncertainty is being actively researched in the field of artificial intelligence and expert systems. Such work—as well as engineering practice in related applications, such as extreme seismic events or nuclear power plant risk assessments—should be reviewed for procedures and experience, including exercises in multiple-expert opinion assessment and aggregation. Experiments that compare model-predicted flows with observed extreme flood flows are needed for valid assessments (see chapters 3 through 5).

3
Flood-Based Statistical Techniques

INTRODUCTION

This chapter considers estimation of extreme flood probabilities by methods involving probabilistic modeling of flood flows and statistical analysis of flood flow data. Modeling here refers to distributional assumptions regarding the underlying random variables; its meaning is elaborated later. Interest is focused on the annual peak discharge at a given site or location in space.

Let Y denote the random variable representing the annual peak discharge (maximum instantaneous flow rate in a water year) at the given site. The distribution of Y is unknown. Two estimation problems are to be considered with emphasis on the second:

(i) Estimate $P[Y > y_0]$ for large fixed y_0,
(ii) Estimate the qth quantile of Y, denoted by ξ_q and defined by $P[Y \leq \xi_q] = q$, for large q.

The first of the two is referred to as *tail probability estimation* and the second as *tail quantile estimation*. Estimation itself can be (a) *point estimation* accompanied by some measure of precision, often in the form of standard error of estimator; or, (b) *interval estimation*, usually in the form of confidence intervals.

For the sake of specificity and brevity, the discussion in this chapter focuses on tail quantile estimation of flood peak discharges. With a few minor modifications in specific details, the same methodologies could be applied to flood volumes or other measures of flood magnitude. Similarly, the estimation of tail probabilities is regarded as fundamentally equivalent to the estimation of tail quantiles. Certainly a probability plot of estimated tail quantiles can be used to read off tail-probability estimates for any specified discharge. It is recognized that the statistical performance of such estimates might not be as desirable as that of estimates specifically derived for probabilities. Nonetheless, the fundamental mathematical structures and the essential points of discussion in this chapter are as applicable to probability estimation as to quantile estimation and it is expected that any necessary adaption of the details for application to probability estimation can be supplied by the interested reader.

Estimation is based on all available flow data pertinent to the site of interest, including systematic flow data at the site as well as at neighboring regional sites and any available historical/paleoflood data. The data are viewed as a value of some random vector, and *modeling* is defined as the making of assumptions about the probabilistic structure of such a random vector. Key components are *regionalization* and inclusion of *historical/paleo information* in the modeling and analysis. Another key component is *robustness*. Loosely speaking, an estimator of the quantile ξ_q is robust if it performs well for a wide variety of underlying population distributions of Y. Robustness is important and deserves careful consideration, inasmuch as it is unlikely that the distribution of Y can be correctly identified.

By its definition, a flood peak is an extreme event, which suggests *extreme-value theory* may have some place in a study of this sort, and it does indeed play two roles, one in the modeling itself and the other in robustness considerations.

A more detailed discussion of the problem, along with some illustrations, appears in the ensuing subsections. No attempt is made to review the literature or to list all contributors. Although any recommended procedure may involve *regionalization*, inclusion of *historical/paleoflood data*, and *robustness*, it is not convenient to discuss all of these jointly.

PROBLEM DELINEATION–BACKGROUND–MODELING–NOTATION

A general mathematical framework, along with the requisite notation, for describing the problem follows. Streamflow at a given point in space varies over time. Let t be the variable that indexes *time* for some convenient time scale and let s be the variable that indexes *space* or *site* location. For example, s could be a two-dimensional vector identifying a point in space (site) by giving its latitude and longitude. The possible values of s may be the points in space that trace a stream. Let $X(t,s)$ denote the streamflow at a particular instant t and site s. The stochastic process $\{X(t,s):t$ ranges over time and s ranges over space$\}$ is the process of interest. There may be some probabilistic structure that governs the behavior of the stochastic process. This structure is unknown, but it is rich enough that it gives the *distribution of the stochastic process, which is defined to be the collection of all possible finite dimensional joint distributions*; a representative finite dimensional joint distribution is the joint distribution of

$$X(t_1, s_1), ..., X(t_k, s_k) \text{ for arbitrary}$$
$$(t_1, s_1), ..., (t_k, s_k).$$

By *model* we mean any set of assumption(s) about the probabilistic structure of the stochastic process. It is recognized and accepted that there is at present no way of identifying the true distribution of the entire stochastic process either from physical or empirical considerations. But, the advantage of such a general mathematical setting is that it is the most general setting, and that it inherently admits *dependence in both time and space*, a feature that is observed in nature.

More specific stochastic structures can be deduced from the general setting. For example, time can be discretized. Define

(3.1) $$Y_j(s) = \sup\{X(t,s)\},$$
$$t \; \epsilon \text{ year } j,$$

which is the peak flow at site s for year j. (Here, sup, short for supremum, can be considered an abbreviation for maximum.) Let s_0 denote the site of interest and set $Y_j = Y_j(s_0)$. Let us restrict our attention to models that say that the distribution of Y_j does not depend on j; this is a type of stationarity called spatial homogeneity (Dalrymple, 1960). Let $F(\cdot)$ be the cumulative distribution function (cdf) of Y_j, i.e.,

(3.2) $$F(y) = F_{Y_j}(y) = P[Y_j \leq y].$$

Of course, the true distribution $F(\cdot)$ is not known and never likely to be completely identifiable. The qth quantile of $F(\cdot)$, denoted by $\xi_q(F)$, is defined as

(3.3) $$\xi_q(F) = \inf_y \{F(y) \geq q\},\ 0 \leq q \leq 1.$$

(Here, inf, short for infimum, can be considered an abbreviation for minimum.) Note that if $F(\cdot)$ is absolutely continuous and ξ_q is a point of increase of $F(\cdot)$ then ξ_q is given by $F(\xi_q) = q$. As mentioned earlier, our interest lies in the estimation of

(i) $\overline{F}(y_0) = 1 - F(y_0) = P[Y_j > y_0]$ for large fixed y_0, or
(ii) $\xi_q(F)$ for large q.

$\xi_q(F)$ is the annual flood (instantaneous peak or some other flood indicator) having return period $T = 1/(1 - q)$; e.g., the .99th quantile is the flood with a 100-year return period. F is included in the notation to stress that the quantile depends on the (unknown) cumulative distribution function, cdf $F(\cdot)$. Sometimes a parametric model is assumed for the distribution of Y_j. That is, it is assumed that the true $F(\cdot)$ belongs to some parametric family (such as the generalized extreme value, log Pearson Type III, Wakeby, log normal, or other distribution), say

(3.4) $$\{F(\cdot;\theta) : \theta \in \Theta\}.$$

Here θ is our *generic parameter* (possibly a vector) and Θ is our *parameter space*. Now $F(\cdot)$ is indexed by θ so one can write

(3.5) $$\xi_q(F) = \xi_q(\theta)$$

and the assumed parametric family gives the function $\xi_q(\cdot)$. For example, if

$$\{F(y;\theta) : \theta \in \Theta\} = \{(1 - e^{-\theta y})I_{(0,\infty)}(y) : \theta > 0\}$$

then $1 - e^{-\theta \xi_q} = q$ implies $\xi_q(\theta) = -\ln(1 - q)/\theta$ for $0 < q < 1$. Here and later $I_{(a,b)}(y)$ is the usual indicator notation, i.e., $I_{(a,b)}(y)$ equals

1 if $a < y < b$ and equals 0 otherwise. Parametric modeling for single site data is considered in the next subsection.

Having introduced the *modeling* aspect of the problem, the *data* aspect comes next. Let $\underline{x}$ be generic notation for all available data pertinent (interpreted to be all data to be used in analysis) to site s_0, the site of interest. In general, $\underline{x}$ will contain flow data at several sites and include both systematically recorded flow data as well as historical/paleodata that may have been transformed into flow rates. It may be that $\underline{x}$ consists only of systematic data or only of historical data. To bring the problem into the realm of statistical theory, $\underline{x}$ is considered a value of some random vector (or random function), say $\underline{X}$. Modeling can now be condensed to making assumptions about the distribution of $\underline{X}$. Based on the data $\underline{x}$, our objective is to estimate (1) the tail probability $F(y_0) = P[Y_j > y_0]$ for fixed large y_0, or, (2) the tail quantile $\xi_q(F)$ for fixed large q.

In the quantile point estimation case we seek a statistic (defined to be some function of the data), say $t(\cdot)$, where $t(\cdot)$ evaluated at $\underline{x}$ gives our point estimate of $\xi_q(F)$. The estimation problem is first to discover or find what statistics $t(\cdot)$ make useful estimators, and then to arrive at some criteria for assessing the goodness of the estimator. There are several classical measures of goodness including low bias, asymptotic unbiasedness, low mean square error, uniformly minimum variance unbiasedness, consistency, efficiency, etc. Consistency is the property that the estimator converges in some sense to the true parameter value when the size of the data set is expanded to infinity. Bias is the difference between the expected value of the estimator and the value of the parameter to be estimated, i.e.

(3.6) $\qquad bias$ of estimator $t(\cdot)$ of $\xi_q(F)$ is $E[t(\underline{X})] - \xi_q(F)$.

An estimator $t(\cdot)$ is *unbiased* if it has bias equal to zero so that

$$E[t(\underline{X})] = \xi_q(F).$$

The *mean square error* of an estimator $t(\cdot)$ of $\xi_q(F)$ is

(3.7) $\qquad \text{MSE}_t = E\{[t(\underline{X}) - \xi_q(F)]^2\}.$

Since MSE measures the spread, in the sense of squared error, of the estimator to be estimated, small MSE is desirable. When an estimator is unbiased the MSE equals the variance of the estimator.

The *standard error of estimate* is the standard deviation of the estimator, a useful measure of the goodness of unbiased (or nearly unbiased) estimators. *Efficiency* of an estimator is a notion that measures how well that estimator does relative to the best that can be done.

Just as there are several standard measures of the goodness of an estimator, there are several standard procedures for finding estimators. These include the method of least squares, method of moments, method of probability weighted moments, maximum likelihood procedure, minimum distance method, and the Bayesian method. Among statisticians, the contemporary standard of these is the method of maximum likelihood estimation (MLE), primarily because of its desirable property of asymptotic efficiency, illustrated below. For small finite samples, typical in flood-frequency analysis, MLE's may be inefficient, and other estimation procedures may be preferred.

To illustrate the method of maximum likelihood let us assume a parametric model for $\underline{X}$, using the same notation as in (3.4), but now it is the distribution of $\underline{X}$ that is being considered. Let $\underline{X}$ have a joint density belonging to a parametric family, say

$$(3.8) \qquad \{f_{\underline{X}}(\underline{x};\theta) = f(\underline{x};\theta) : \theta \in \Theta\},$$

where both the function $f(\cdot;\cdot)$ and Θ are known but the value of θ is unknown; θ may be a vector. The MLE of θ, denoted by $\hat{\theta}$, is obtained by maximizing the likelihood function as a function of θ for fixed $\underline{x}$ where the likelihood function is the joint density of $\underline{X}$. By the invariance property of maximum likelihood estimation, the MLE of $\tau(\theta)$ for a known function $\tau(\cdot)$ is $\tau(\hat{\theta})$; that is, a function of a MLE is itself a MLE. In our case $\tau(\theta) = \xi_q(\theta)$ and the function $\xi_q(\cdot)$ will be derived and known from the assumed model. Asymptotic theory for MLEs exists and it enables one (under regularity conditions) to obtain an approximate standard error of an estimator, or an approximate confidence interval. Also this same theory says that maximum likelihood estimation leads to asymptotically efficient estimators. For later use, the asymptotic theory alluded to above is: Under regularity conditions,

$$(3.9) \qquad \theta \stackrel{.}{\sim} \text{MVN}[\theta, \mathbf{I}_n^{-1}(\theta)],$$

where $\mathbf{I}_n(\theta)$ is the Fisher information matrix having ijth entry

$$E_\theta \left[\frac{\partial \log f(\underline{X};\theta)}{\partial \theta_i} \frac{\partial \log f(\underline{X};\theta)}{\partial \theta_j} \right],$$

$\dot\sim$ reads "is asymptotically distributed as," and MVN$[\theta,\mathbf{I}_n^{-1}(\theta)]$ abbreviates multivariate normal with mean vector θ and variance-covariance matrix $\mathbf{I}_n^{-1}(\theta)$. Here $\theta = (\theta_1,...,\theta_k)$ is a k-dimensional vector. If $\tau(\theta) = [\tau_1(\theta),...,\tau_r(\theta)]$ and

$$\tau = \left[\frac{\partial \tau_i(\theta)}{\partial \theta_j} \right]_{r \times k}$$

then

(3.10) $\qquad \tau(\theta) \dot\sim$ MVN $[\tau(\theta), \tau \mathbf{I}_n^{-1} \tau']$.

The subscript n is a sample size indicator. Use of these results germane to our problem appears in subsections below.

One valuable feature of ML estimation is that the ML estimate and its asymptotic variance can be obtained by adjusting the model, so that an explicit formula for the estimator is not required. Hence for a given data set one can use one of many available computer optimization routines to maximize the likelihood and possibly some numerical routine to evaluate the Fisher information matrix to get the standard error of the estimator. Essentially, once a parametric model has been selected, ML estimation says you have done about as well as you can do asymptotically.

How does one pick and verify the model? One possibility is to perform some sort of goodness of fit based on the data. This is not likely to be a viable option inasmuch as there is rarely enough data to discriminate among competing parametric models. Another possibility is to derive the model based on other considerations, including regionalization or physical considerations. Within available statistical theory, *extreme-value theory* has the possibility of assisting one in building a model. We sketch some particulars associated with *extreme-value theory* next. Leadbetter et al. (1983) give a detailed treatment of this theory.

Let $Z_1, Z_2,...,Z_n$ be a sequence of independent and identically distributed (iid) random variables with common cdf $F(\cdot)$. Let $M_n = \max(Z_1,...,Z_n)$. Now the exact cdf of M_n is given by:

$$F_{M_n}(m) = P[M_n \leq m] = F^n(m).$$

Further, according to the *extremal types theorem*, if there exist sequences of constants, say $\{a_n\}$ and $\{b_n\} > 0$ such that

$$P[(M_n - a_n)/b_n < x] = F^n(b_n x + a_n) \to \Lambda(x), \tag{3.11}$$

where $\Lambda(\cdot)$ is a nondegenerate* limit cdf, then $\Lambda(\cdot)$ must be one of the three extreme value distributions given by:

$$\text{Type I}: \Lambda_1(x) = \exp(-e^{-x})$$
$$\text{Type II}: \Lambda_2(x;\gamma) = \exp(-x^{-\gamma})I_{(0,\infty)}(x), \text{ where } \gamma > 0$$
$$\text{Type III}: \Lambda_3(x;\gamma) = \exp[-(-x)^\gamma]I_{(-\infty,0)}(x) + I_{(0,\infty)}(x),$$
$$\text{where } \gamma > 0.$$

The latter two types are parametric families with parameter γ. We say $F(\cdot)$ is *attracted* to $\Lambda(\cdot)$, or belongs to the *domain of attraction* of $\Lambda(\cdot)$, if (3.11) is satisfied, and we write $F \in D(\cdot)$. Examples of distributions $F \in D(\Lambda_1)$ are: exponential, gamma, Weibull, normal, log normal, logistic, and Λ_1 itself. Examples of distributions $F \in D(\Lambda_2)$ are: t-distribution, Pareto, Cauchy, log gamma, and Λ_2 itself. Finally, examples of $F \in D(\Lambda_3)$ are: uniform, beta, and Λ_3. Although not all distributions have a domain of attraction, most do and one sees the potential of using extreme-value theory in modeling. In fact the iid assumption of the extremal types theorem can be relaxed (see Leadbetter et al., 1983) and hence one may be able to justify assuming that the annual peak discharge random variable Y (it is the largest of several discharges) has an approximate extreme-value distribution. The convergence to a limiting extreme-value distribution can be quite slow, which affects the usefulness of the approximation. Since the actual type of extreme-value distribution may be unknown, a generalized extreme-value (GEV) distribution, one that is rich enough in parameters that it contains all three of the types, seems useful. Such a distribution does exist and can be given by

$$\Lambda(x;\alpha,\beta,\kappa) = \exp\{-[1 - \kappa\left(\frac{x-\alpha}{\beta}\right)]^{1/\kappa}\} \text{ for}$$
$$\kappa(x-\alpha)/\beta < 1, \text{ where } \beta > 0. \tag{3.12}$$

---
*A degenerate cdf has all its mass at a single point.

The parameter κ doubles as the type selector and the replacement for γ. $\kappa = 0$ [defined as $\kappa \to 0$ in (3.12)] corresponds to Type I, $\kappa < 0$ corresponds to Type II, and $\kappa > 0$ to Type III. α and β are location and scale parameters, respectively. We will return to the GEV in the next subsection.

As mentioned earlier, an estimator, say $t(\cdot)$, of $\xi_q(F)$ is *robust* if it performs well for a wide variety of cdf's $F(\cdot)$ of Y. Robustness can be quantified in a number of ways. One way is to utilize the notion of efficiency. Efficiency too can be defined in a variety of ways; for example, efficiency of an estimator can be defined to be the reciprocal of the variance of the estimator times a best lower bound for such variances. Suppose one is interested in comparing estimators $t_1(\cdot)$ and $t_2(\cdot)$ to ascertain which one is the more robust and suppose robustness is to be judged relative to some family, say $\mathcal{F}$, of F's. Let eff(t,F) be the *efficiency of estimator t* when F is the correct cdf. If eff$(t,F) = 1$ for some $t(\cdot)$ and $F(\cdot)$ then estimator $t(\cdot)$ is best (most efficient) when $F(\cdot)$ is the cdf. Efficiencies near one are desired. If eff$(t_1,F) \geq$ eff(t_2,F) for all $F \in \mathcal{F}$ then naturally one prefers t_1 and agrees t_1 is more robust than t_2. What is likely to occur is that t_1 will be more efficient than t_2 for some F and less efficient for other F, making an unambiguous comparison impossible. In such cases one could use a maximin concept and prefer t_1 over t_2 if

$$(3.13) \qquad \inf_{F \in \mathcal{F}} \text{eff}(t_1, F) \geq \inf_{F \in \mathcal{F}} \text{eff}(t_2, F),$$

since then t_1's worst efficiency relative to F is higher than t_2's worst efficiency. One would like that there be an estimator, say $t^*(\cdot)$, that has "high" efficiency for all $F \in \mathcal{F}$.

Probability Weighted Moments

A method of estimation based on probability weighted moments (PWMs) has recently been suggested as an alternative to the method of maximum likelihood, or other methods. As is the case in the classical method of moments estimation, estimators are obtained by equating sample PWMs to population PWMs which are defined as follows. [An extensive discussion of PWMs is given by Hosking (1986a).]

Let X be a real-valued random variable with cumulative distribution function $F(\cdot)$. The (population) PWMs of X or F are defined to be

(3.14) $$\mu_{k,r,s} = E\{X^k[F(X)]^r[1-F(X)]^s\},$$

where k, r, and s are real numbers. If r and s are nonnegative integers and $X_{j:m}$ is the jth smallest order statistic of a random sample of size m from the distribution F then

(3.15) $$\mu_{k,r,s} = \frac{r!s!}{(r+s+1)!} E[X^k_{r+1:r+s+1}].$$

This linkage between expectations of order statistics and PWMs suggests taking appropriate linear combinations of order statistics as sample PWMs and this is what is done. Define

(3.16) $\alpha_r = \mu_{1,0,r}$ and $\beta_r = \mu_{1,r,0}$ for $r = 0, 1, ...$ Further define

(3.17) $$A_r = \left\{1 \Big/ \left[m \binom{m-1}{r}\right]\right\} \sum_{i=1}^{m} \binom{m-i}{r} X_{i:m}, \text{ and}$$

(3.18) $$B_r = \left\{1 \Big/ \left[m \binom{m-1}{r}\right]\right\} \sum_{i=1}^{m} \binom{i-1}{r} X_{i:m} \text{ for } r = 0, 1, ..., m-1.$$

The A_r's and B_r's are called sample PWMs. It is easy to show that

$$E[A_r] = \alpha_r \text{ and } E[B_r] = \beta_r,$$

so that A_r and B_r are unbiased estimators of α_r and β_r, respectively. In practice the combinatorial coefficients in Equations (3.17) and/or (3.18) are sometimes replaced by asymptotic equivalents.

Estimators of, say, k parameters can be obtained by equating k sample PWMs to k population PWMs and solving the equations for the parameters.

Since estimators obtained by the method of PWMs will be functions of linear combinations of order statistics, an asymptotic distribution for a linear combination of order statistics is useful. A simple result follows.

As before, let $X_{1:m} \leq ... \leq X_{m:m}$ be the order statistics corresponding to a random sample of size m from F. Let $F_m(\cdot)$ be the sample cdf, then, under mild conditions on $J(\cdot)$, a weight function, and $F(\cdot)$,

$$\text{(3.19)} \qquad m^{1/2}\{T_m - \int_{-\infty}^{\infty} xJ[F(x)]dF(x)\},$$

converges in distribution to a normal with mean 0 and variance σ^2, where

$$T_m = (1/m)\sum_{j=1}^{m} J(j/m)X_{j:m} = \int_{-\infty}^{\infty} xJ[F_m(x)]dF_m(x)$$

and

$$\sigma^2 = 2\int\int_{x<y} J[F(x)]J[F(y)]F(x)[1-F(y)]dxdy.$$

Result (3.19) is useful in finding the asymptotic standard error of a PWM estimator.

Moments of Floods and Rainfall Maxima

The moments of a random variable are useful characteristics of the random variable. In this section we discuss (population) moments and sample moments as estimators of population moments, concluding with some comments regarding data.

Four characteristics of a random variable X based on the first four moments are:

$$\text{the mean } \mu = \mu_X = E(X)$$
$$\text{the standard deviation } \sigma = \sigma_X = \{E[X-\mu]^2\}^{0.5},$$
$$\text{the skewness coefficient } \gamma = \gamma_X = \{E[X-\mu]^3\}/\sigma^3,$$
$$\text{the kurtosis coefficient } \lambda = \lambda_X = \{E[X-\mu]^4/\sigma^4\}.$$

The moment estimators for a sample of size n are

$$\text{(3.20)} \qquad \hat{\mu} = \sum_{i=1}^{n} X_i/n,$$

$$(3.21) \quad \hat{\sigma} = \left[\sum_{i=1}^{n} X_i^2/n - \hat{\mu}^2\right]^{0.5},$$

$$(3.22) \quad \hat{\gamma} = \left[\sum_{i=1}^{n} X_i^3/n - 3\hat{\mu}\hat{\sigma}^2 - \hat{\mu}^3\right]/\hat{\sigma}^3,$$

$$(3.23) \quad \hat{\lambda} = \left[\sum_{i=1}^{n} X_i^4/n - 4\hat{\mu}\hat{\sigma}^3\hat{\gamma} - 6\hat{\mu}^2\hat{\sigma}^2 - \hat{\mu}^4\right]/\hat{\sigma}^4.$$

Equation (3.20) is an unbiased estimate of μ, but the succeeding estimators are increasingly biased. The magnitude of the bias in equations (3.21)–(3.23) is dependent upon n and the distribution of X. For the small n's likely to be encountered in sequences of flood maxima these biases may be quite large.

Consider the results shown in Figure 1 where the cumulative distributions of $\hat{\mu}$, $\hat{\sigma}$, and $\hat{\gamma}$ have been plotted and their average values shown for comparison purposes. The figure shows that for log normal distribution with $\mu = 0$, $\sigma = 1$, and $\gamma = 15$ (indicative of an extreme skew case) the biases in the higher moments can be expected to be considerable. In this case, for an $n = 30$, $E[\hat{\sigma}]$ is 0.77, while $E[\hat{\gamma}]$ is 3.6. Regionalization procedures that depend upon averaging at-site estimates of the moment estimators can give poor quantile estimates in the upper tails that are of interest in flood studies. The problems of small sample biases and nonnormal distributions for estimates of higher moments (Hosking, 1986a). Schaefer (1987) has reported a successful application of this concept with rainfall maxima for the state of Washington.

It is an algebraic fact that the value of the sample skew coefficient is bounded as follows (Kirby, 1974):

$$(3.24) \quad \hat{\gamma} \leq \frac{n-2}{\sqrt{n-1}}.$$

Equation (3.24) is distribution free, and for an n of 30 the upper bound of $\hat{\gamma}$ is 5.2, which is evident in Figure 1.

Figure 2 shows that hourly rainfall maxima tend to be highly skewed, and with skew-kurtosis relationships that are quite GEV-like. As might be expected, 24-hour point rainfall data are less variable and less skewed than the corresponding hourly data, but still quite GEV-like in their skew-kurtosis relationship (Figure 2). Similar results have been observed for point rainfall data for the

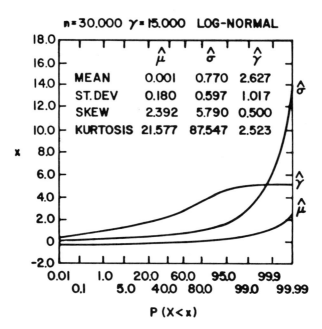

FIGURE 1 Cumulative distributions of $\hat{\mu}$, $\hat{\sigma}$, and $\hat{\gamma}$ for a three-parameter log normal distribution with $\mu = 0$, $\sigma = 1$, and $\gamma = 15$ for $n = 30$ (after Wallis et al., 1974).

state of Washington, as well as for European and New Zealand rainfall data (Wallis, personal communication, 1987).

Covariance Among Flood and Rainfall Maxima

A final item of background is that of covariance and correlation. The covariance between two random variables X and Y is

$$\mathrm{Cov}(X,Y) = E\{[X - \mu_X][Y - \mu_Y]\}.$$

The standardized measure of the linear relationship between X and Y is the coefficient of correlation

$$\rho = \rho_{X,Y} = \frac{\mathrm{Cov}(X,Y)}{\sigma_X \sigma_Y}. \tag{3.25}$$

ρ can be estimated from n pairs of observations (X_1, Y_1), $(X_2, Y_2),...,(X_n, Y_n)$ by the sample correlation coefficient defined by

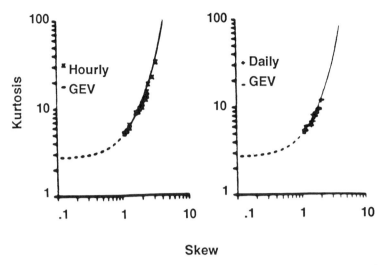

FIGURE 2 Skew-kurtosis relationship for annual maxima of California point rainfall data estimated by the WAK/R algorithm and with no small sample bias corrections attempted (after Wallis, 1982).

$$(3.26) \quad \hat{\rho} = \frac{\sum_{i=1}^{n}(X_i - \hat{\mu}_x)(Y_i - \hat{\mu}_y)}{\left[\sum_{i=1}^{n}(X_i - \hat{\mu}_x)^2 \sum_{i=1}^{n}(Y_i - \hat{\mu}_y)^2\right]^{0.5}}.$$

Benson (1962) made a study of 164 New England annual flood records in which the average correlation between sites was reported to be 0.26. Using an equation attributed to Yule and Alexander,

$$(3.27) \quad N_e = \frac{N}{1 + (N-1)\rho},$$

where N is the number of sites in the region, ρ is the correlation between sites, and N_e is the equivalent number of independent sites, Benson stated that the data from all 164 sites of his study were equivalent to that of only three independent sites. This means that the standard errors of regionalized estimates (of the mean) would be reduced by a factor of about 2 compared to the at-site estimates, rather than by a factor of about 12. Under an apparent assumption that regionalized quantile estimates would be at least as subject to error as regionalized means, Benson concludes from this finding that it is not possible to reduce sampling errors indefinitely simply by adding more stations to a regional gaging network. In Benson's example, 4 sites would provide nearly the same accuracy (for a regional mean) as 164 sites. Benson thus concludes that operation of a gage network is more beneficial for developing information about floods at ungaged sites than for reducing estimation error at gaged sites. More recent and explicit studies concerning the effects of cross-correlation on accuracy of regionalized quantile estimates are developed in example 3.8 later in this chapter. [See also Stedinger (1983).]

Rain gages are small and usually widely spaced in relationship to the moving cells of intense precipitation that give rise to the annual maxima; hence short-time-interval rain gage maxima exhibit little or no cross-correlation between gages.

SINGLE SITE ESTIMATION

With the background of the previous section, we turn now to the single site frequency estimation, the simplest statistical situation that can be encountered. The primary assumption throughout this subsection is that the only data to be analyzed are data at the site of interest. Such data may include historical/paleoflood data. The purpose here is to give some indication of how well one can estimate quantiles in some simple and idealized settings. The examples show rather dramatically that large errors can be expected when the probabilities of rare floods are estimated with only the data from the site of interest.

Simple Parametric Setting

Suppose now that the data x consists solely of n years worth of annual peak flows at the site of interest. Thus $\underline{x} = (y_1,...,y_n)$ where y_i is the actual peak flow for year i, $i = 1,...,n$. y_i is a value of random variable Y_i.

Example 3.1 Assume $Y_1,...,Y_n$ are independent and identically distributed with common exponential distribution with rate parameter θ. That is, $Y_1,...,Y_n$ are independent and identically distributed as $f(y;\theta) = \theta e^{-\theta y}$ for $y \geq 0$. We chose the exponential distribution for demonstration purposes only; its properties are well known, allowing conclusions to be drawn analytically and simply.

Let $\xi_q(\theta)$ be the qth quantile. Based on $Y_1,...,Y_n$ we want to estimate $\xi_q = \xi_q(\theta)$. Since q = F(ξ_q) = $1 - e^{-\theta \xi_q}$, $\xi_q(\theta) = -\ell n \, p/\theta$

where $p = 1 - q$. Now, it is known that the MLE of θ is $1/\bar{Y}$ so the MLE of $\xi_q(\theta)$ is $(-\ell n\ p)\bar{Y}$. Here the exact distribution of the MLE is known (it's a gamma) but the central limit theorem tells us

$$(-\ell n\ p)\bar{Y} \stackrel{.}{\sim} N[\xi_q(\theta), (-\ell n\ p)^2/n\theta^2]$$
$$= N[\xi_q(\theta), \xi_q^2(\theta)/n].$$

Therefore, the approximate standard error of the MLE is $\xi_q(\theta)/\sqrt{n}$. Also, since $\sqrt{n}[(-\ell n\ p)Y - \xi_q(\theta)]/\xi_q(\theta) \stackrel{.}{\sim} N(0,1)$ an approximate $100(1-\alpha)$ percent confidence interval estimator of $\xi_q(\theta)$ is

$$\frac{(-\ell n\ p)\bar{Y}}{1 + z/\sqrt{n}}, \frac{(-\ell n\ p)\bar{Y}}{1 - z/\sqrt{n}},$$

where $\Phi(z) - \Phi(-z) = 1 - \alpha$. The length of this confidence interval estimator is

$$(-\ell n\ p)\bar{Y}\frac{2z/\sqrt{n}}{1 - z^2/n}.$$

Since both the standard error of estimator and length of confidence interval are proportional to $(-\ell n\ p)$, one can easily see the effect of p on the precision of estimation.

$\dfrac{p}{-\ell n\ p}$ ‖	$\dfrac{1/100}{4.6}$	$\dfrac{1/1000}{6.9}$	$\dfrac{1/10,000}{9.2}$	$\dfrac{1,000,000}{11.5}$.

Under this model the length of the confidence interval estimator for the 10,000-year flood is twice what it is for the 100-year flood.

As a second example consider the case of the log normal distribution.

Example 3.2 Assume $Y_1,...,Y_n$ are independent and identically distributed as log normal with parameters μ and σ^2. Equivalently, assume $\ell n\ Y_1,...,\ell n\ Y_n$ are independent and identically distributed as $N(\mu,\sigma^2)$. Note

$$q = P[X_1 \leq \xi_q(\theta)] = P[\ell n\ X_1 \leq \ell n\ \xi_q(\theta)] = \Phi\left(\frac{\ell n\ \xi_q(\theta) - \mu}{\sigma}\right),$$

which implies that $\ell n\ \xi_q(\theta) = \mu + z_q\sigma$ where z_q is given by $\Phi(z_q) = q$. Finally $\xi_q(\theta) = \exp\{\mu + z_q\sigma\}$. In terms of our generic $\theta, \theta = (\mu,\sigma)$ and $\tau(\theta) = \xi_q(\theta) = \exp\{\mu + z_q\sigma\}$. The MLE of θ is routinely obtained and is $(\Sigma \ln Y_i/n, [\Sigma(\ln Y_i - \overline{\ln Y})^2/n]^{1/2})$ from which the MLE of $\xi_q(\theta)$ can be seen to be

$$\exp\{\overline{\ln Y} + z_q[\Sigma(\ln Y_i - \overline{\ln Y})^2/n]^{1/2}\}.$$

Further, the asymptotic distribution [see (3.10)] of this latter estimator is $N[\xi_q, (\partial \xi_q/\partial \mu, \partial \xi_q/\partial \sigma) \mathbf{I}_n^{-1}(\theta) (\partial \xi_q/\partial \mu, \partial \xi_q/\partial \sigma)']$ and the entries can be computed and a comparison can be made. Note that $\partial \xi_q/\partial \mu = \xi_q$ and $\partial \xi_q/\partial \sigma = z_q \xi_q$ so that the asymptotic variance becomes

$$(\xi_q, z_q \xi_q) \mathbf{I}_n^{-1}(\theta)(\xi_q, z_q \xi_q)'.$$

Now,

$$f(Y;\theta) = \frac{1}{Y\sqrt{2\pi}\sigma} \exp\left[\frac{-1}{2\sigma^2}(Y-\mu)^2\right], \text{ so}$$

$$\ln f(Y;\theta) = -\ln Y\sqrt{2\pi} - \ln \sigma - \frac{1}{2\sigma^2}(Y-\mu)^2,$$

$$\frac{\partial}{\partial \mu} \ln f(Y,\theta) = \frac{(Y-\mu)}{\sigma^2} \text{ and } \frac{\partial}{\partial \sigma} \ln f(Y;\theta) = -\frac{1}{\sigma} + \frac{(Y-\mu)^2}{\sigma^3},$$

$$E\left\{\left[\frac{\partial}{\partial \mu} \ln f(Y;\theta)\right]^2\right\} = E(Y-\mu)^2/\sigma^4 = \sigma^{-2},$$

$$E\left\{\left[\frac{\partial}{\partial \sigma} \ln f(Y;\theta)\right]^2\right\} = \left\{\left[\frac{(Y-\mu)^2}{\sigma^3} - \frac{1}{\sigma}\right]^2\right\}$$

$$= \frac{1}{\sigma^6} E\{[(Y-\mu)^2 - \sigma^2]^2\} = \sigma^{-6}[E(Y-\mu)^4$$

$$- 2\sigma^2 E(Y-\mu)^2 + \sigma^4] = 2\sigma^{-2}, \text{ and}$$

$$E\left\{\left[\frac{\partial}{\partial \mu} \ln f(Y;\theta)\right]\left[\frac{\partial}{\partial \sigma} \ln f(Y;\theta)\right]\right\} = 0.$$

Therefore

$$\mathbf{I}(\theta) = \begin{bmatrix} n/\sigma^2 & 0 \\ 0 & 2n/\sigma^2 \end{bmatrix}$$

and

$$\mathbf{I}^{-1}(\theta) = (1/n)\begin{bmatrix} \sigma^2 & 0 \\ 0 & \sigma^2/2 \end{bmatrix}.$$

Finally, the asymptotic variance is

$$(\xi_q, z_q\xi_q)\mathbf{I}^{-1}(\theta)(\xi_q, z_q\xi)' = \frac{\sigma^2 \xi_q^2}{n}(1+z_q^2/2).$$

Thus the asymptotic standard error of estimate is

$$\sigma \exp\{\mu + z_q\sigma\}\left(\frac{1}{n} + \frac{z_q^2}{2n}\right)^{1/2},$$

which is proportional to $\exp\{\sigma z_q\}(1+z_q^2/2)^{1/2} = g(p)$, which is computed in the table below for $\sigma = 1$ and three values of p.

$$\frac{p}{g(p)} \parallel \frac{1/100}{19.7} \mid \frac{1/1000}{52.8} \mid \frac{1/10,000}{331.6} \mid .$$

In this case, the length of the confidence interval estimators for the 10,000-year flood is almost seventeen times what it is for the 100-year flood.

Example 3.3 As mentioned in the previous subsection, the GEV is a candidate distribution. Suppose $Y_1, \ldots, Y_n$ are independent and identically distributed with common cdf given by (3.12). Now $\theta = (\alpha, \beta, \kappa)$ and $\xi_q = \xi_q(\theta) = \alpha + (\beta/\kappa)[1 - (-\ln q)^\kappa]$, where $\xi_q(\theta)$ is defined at $\kappa = 0$ as the limit as $\kappa \to 0$. A scheme for finding the MLE of θ and then the MLE of $\xi_q(\theta)$ by invariance is given in Prescott and Walden (1980, 1983). The Fisher information matrix is also given and the requisite regularity conditions are satisfied for $\kappa < 1/2$ so one can obtain an asymptotic standard error of estimator. Hosking et al. (1985a) propose estimating the parameters and quantiles of the GEV by the method of probability weighted moments. Asymptotic properties of the estimators were investigated for small and moderate samples via simulation. The method of probability-weighted moments compares favorably with the method of maximum likelihood. Several tables and graphs are given that portray efficiencies. Such analysis is not repeated here; the main point is that for this parametric family, the analytics necessary to see how well one can do have been accomplished.

Tail Behavior Results

In all three of the above examples the data record consisted of n years of annual peak flow data. Such peak flows may have had different causes. For example, peak flows could result from just snowmelt, thunderstorm and snow melt, thunderstorm on frozen ground, thunderstorm on ground saturated by a general wet period, or hurricane and frontal produced rainfall. One can argue that each of these peak types ought to have its own distribution and that the true annual peak discharge distribution is a mixture of the distributions associated with the respective causes (Hazen, 1930). It is unlikely that one would ever have sufficient data to characterize either the peak type distributions or the relative weights in the mixture to be given to these types. Some other method is needed. The tail behavior of a mixture is often dictated by the tail behavior corresponding to the distribution in the mixture having the heaviest tail. Thus, one way of circumventing the problem is to consider techniques based on tail behavior and the largest data values. The first of these methods is usually credited to Weissman (1978).

Example 3.4 Assume $Y_1,...,Y_n$ are independent and identically distributed as F and $F \in D(\Lambda_1)$. Let $Y_{1n} \geq Y_{2n} \geq ... \geq Y_{nn}$ be the ordered $Y_1,...,Y_n$, indexed from largest to smallest. For fixed k, there exists $\{a_n\}$ and $\{b_n > 0\}$

(3.28) $$[(Y_{1n} - a_n)/b_n, ..., (Y_{kn} - a_n)/b_n] \to M_k,$$

where M_k is a random vector having density

(3.29) $$g(y_1, ..., y_k) = \exp[-\exp(-y_k) - \sum_{j=1}^{k} y_i] \text{ for } y_1 \geq y_2 \geq ... \geq y_k.$$

Hence $Y_{1n},...,Y_{kn}$ has approximate density given by

(3.30) $$b_n^{-k} g[(y_1 - a_n)/b_n, ..., (y_k - a_n)/b_n].$$

Now MLEs of a_n and b_n, based on $Y_{1n},...,Y_{kn}$ and approximate density (3.30), are

$$\hat{b}_n = \left[\left(\sum_{j=1}^{k} Y_{jn}/k\right) - Y_{kn}\right] \text{ and } \hat{a}_n = \hat{b}_n \ln k + Y_{kn}.$$

By assumption, $F^n(b_n z + a_n) \approx \exp\{-\exp(-z)\}$, or $F^n(x) \approx \exp\{-\exp[-(x - a_n)/b_n]\}$; so $q^n = F^n(\xi_q) \approx \exp\{-\exp[-(\xi_q - a_n)/b_n]\}$ and ξ_q can be solved for in terms of a_n and b_n and then an approximate MLE ξ_q can be obtained. Further, using the asymptotic theory for MLE one can get an approximate standard error of the estimator. Boos (1984) analyzes the performance of such estimators.

The procedure described in Example 3.4 presumably would be reasonably robust for all $F \in D(\Lambda_1)$. Similar results are available for the other two extreme value types and presumably would also then work for the GEV.

There are other estimation procedures that are based on upper tail behavior. Davis and Resnick (1984) give an estimator of a tail probability based on the upper k order statistics that is consistent ($k \to \infty$, but $k/n \to 0$ as $n \to \infty$) for a wide class of distributions. Their method is based on the Pareto-like tail behavior that comes out of extreme-value theory. [See Pickands (1975).] Also, their tail probability estimator can be inverted to give a tail quantile estimator.

R. L. Smith (1985) also utilized the Pickands (1975) result to get upper tail behavior of a generalized Pareto-type distribution and ties it to the "peaks over threshold" methods. J. A. Smith (1986) is the most recent contributor along these lines.

A possible advantage of techniques based on tail behavior is that only the "upper" part of the data is used for estimating upper tail probabilities or upper tail quantiles. In doing so one does not have to worry about whether or not the "lower" data values really follow the distribution. A disadvantage, of course, is that these techniques are based on asymptotics and one needs to check whether or not one can invoke asymptotics for the small sample sizes one is apt to encounter in practice.

Nonparametric Procedures

In a certain sense the methods referenced in the last section are nonparametric. Breiman and Stone (1985) give a method for computing estimates and confidence bounds for tail quantiles based on the upper tail of the data that does not depend on an assumed parametric model for the distribution of the data. They essentially fit a quadratic tail model to the upper part of the data and extrapolate. A simulation study with large samples shows the method may have promise for a family of distributions that are neither too-heavy nor too-light tailed.

Incorporation of Historical Data

Historical data includes any information/data on floods that occurred before or after systematic streamflow gauging. Such information/data can be variable in form and/or accuracy. Historical data might be simply the knowledge that a certain water level had been exceeded sometime prior to or after the systematic gauging, or it might be quite specific, such as knowing the flood level and year of such level for several floods. The incorporation of historical data into our modeling depends on the form of data and will be illustrated with two examples.

Example 3.5 Let $Y_1,...,Y_n$ represent the systematic record of annual peak flows for n years. Additionally, suppose that it is known that level y_0 was exceeded exactly k times during a time span of m years. Model by assuming the $Y_1,...,Y_n$ are independent and identically distributed as $f(\cdot;\theta)$, $\theta \in \Theta$ and that $Z_1,...,Z_m$ are indicator variables independently and identically distributed as Bernoulli $\{p = P[Y_i > y_0] = 1 - F(y_0;\theta)\}$. Z_j equals 1 if level y_0 was exceeded in year j and 0 if not for the m years in the historical record. The Y_i's and Z_j's are independent, and $F(\cdot;\theta)$ is the cdf corresponding to density $f(\cdot;\theta)$. The likelihood function becomes

$$(3.31) \qquad \prod_{i=1}^{n} f(y_i;\theta) \prod_{j=1}^{m} [1 - F(y_0;\theta)]^{z_j} [F(y_0;\theta)]^{1-z_j},$$

where $y_1,...,y_n$ represent the observed values of the annual peaks Y_i of the systematic record and $z_1,...,z_m$ are observed values of the indicator variables Z_j. Since level y_0 was exceeded k times during the time span of the historical record,

$$\sum_{j=1}^{m} z_j = k$$

and the likelihood could be rewritten as

$$(3.32) \qquad \left[\prod_{i=1}^{n} f(y_i;\theta)\right] [1 - F(y_0;\theta)]^k [F(y_0;\theta)]^{m-k},$$

which can be maximized as a function of θ, to get the MLE of θ, for any assumed parametric family. Under this model the historical information is not very precise. All that is needed is whether or not a fixed level was exceeded for m years and the number of times that it was. Further, the MLE of a tail quantile can be readily obtained as well as an approximate standard error estimator using the asymptotic theory for MLE's which ought to be quite good in this case. Stedinger and Cohn (1985, 1986a,b) discuss the model assumed here and include some numerical results for the two-parameter log normal model, which indicate what gain can be anticipated by incorporating error-free historical information data into the analysis. They argue that the value of the historical data is greater in higher dimensional parameter models than in lower dimensional ones. Other families of probability distributions ought to be considered; for example, it would be useful to carry out similar analytics for the GEV.

Example 3.6 Assume as in the previous example that $Y_1,...,Y_n$ represent the systematic record of annual peak flows for n years. Additionally, assume that the flood level of each of the k largest floods over a time span of m years is known. (For the model to be assumed here, the years in which such floods occur carry no additional information.) Let $Z_1 \geq \geq Z_k$ denote the flood levels of the k historical floods. Now model by assuming $Y_1,...,Y_n$ are independent and identically distributed as $f(\cdot;\theta)$, $\theta \in \Theta$ and $Z_1,...,Z_k$ are the k largest order statistics of a random sample of size m from $f(\cdot;\theta)$, and that the Y_i's are independent of the Z_j's. The likelihood function is proportional to

$$(3.33) \qquad \prod_{i=1}^{n} f(y_i;\theta)[F(z_k;\theta)]^{m-k} \prod_{j=1}^{k} f(z_j;\theta),$$

where $y_1,...,y_n$ are the values of $Y_1,...,Y_n$ and $z_1 \geq ... \geq z_k$ the values of $Z_1 \geq ... \geq Z_k$. Again the MLE of θ can be found, and transformed into the MLE of a tail quantile, for any assumed parametric model. The actual maximization may have to be accomplished numerically. Again, the asymptotics of maximum likelihood estimation would allow for the computation of an approximate standard error of estimator and the gain earned by including the historical information obtained for any assumed parametric model. One can anticipate slightly greater gain under this model than under that of the first example inasmuch as the information incorporated here is more precise.

Stedinger and Cohn (1986a) argue that it is often more reasonable to assume that the historical data is in the form of those extreme flood levels exceeding a fixed threshold, rather than the fixed number, k, of historical extremes considered in Example 3.6. Such an assumed model is amenable to maximum likelihood analysis similar to that in Example 3.6.

Many other models that incorporate historical information/data into the analysis could be discussed. For instance, the historical information might be such that the level is known for some historic floods and only the exceedance level for others. Both of these types of information can be handled by the method of maximum likelihood. It could be that the actual level or estimate of the historic flood is less precise than that in the systematic record and that one has to account for this uncertainty in the modeling. This could be accomplished by modeling an error component into the historic readings. Or, one could perform a sensitivity analysis to see the effect of changing the magnitudes of the historic data.

It would seem that historic data are ideally suited and amenable to the types of analysis on tail behavior described in the Tail Behavior subsection. After all, the data represent upper extremes and the time span of the historic record ought to be long enough so that the asymptotics of that theory can be reasonably assumed to be valid. On the other hand, if the time span is long, one has to ask whether stationarity is a valid assumption. Such factors as melting of glaciers, erosion, climatic change, and land use changes could make the assumption of stationarity questionable. However, since the assumption of independence of the historic random variables and the systematic ones is certainly viable, the two types of information need not be analyzed the same way. In any case, historical information is potentially quite valuable since it usually is data regarding the tail quantity of interest.

MULTISITE/REGIONAL ANALYSIS

The assumption here is that our data $\underline{x}$ consists of peak flow readings for several sites within a region, but we are interested in estimating quantiles for a particular site. For concreteness and simplicity in initiating the discussion, assume that we have k sites, each site having annual peak flow data for the same n years. Our data can then be displayed as

$$x_{11}, x_{12}, ..., x_{1n}$$

$$x_{21}, x_{22}, ..., x_{2n}$$

$$...$$

$$x_{k1}, x_{k2}, ..., x_{kn}$$

where x_{ij} is the reading for site i and year j. Again, assume x_{ij} is a value of random variable X_{ij}. Modeling now consists of assuming something about the joint distribution of the X_{ij}, $i = 1,...,k$ and $j = 1,...,n$.

Suppose we are interested in estimating quantiles at site 1. From the modeled joint distribution of the X_{ij} the marginal distribution at site 1 can be found and then the desired quantile can be obtained. Let $\underline{X}$ be the vector of X_{ij}. Specifically let $\underline{X} = (\underline{X}_{k1},...,\underline{X}_{kn})$ where $\underline{X}_{kj} = (X_{1j}, X_{2j},...,X_{kj})$, the random variables for the readings at the k sites for year j. A parametric model is one where

$$f_{\underline{X}}(\underline{x}; \theta) = f(\underline{x}; \theta), \theta \in \Theta.$$

for $f(\cdot;\cdot)$ and Θ is assumed known and the parameter θ unknown. θ is likely to be a vector including the parameters indigenous to each of the individual sites.

One might wonder how it is that data at sites other than site 1, the site of interest, can be useful in estimation at site 1. In fact, there are times when neighboring sites are of no use at site 1. If, for instance, all sites were perfectly correlated, there would be no additional information in the neighboring sites, unless of course they are for longer periods. Also, if the site random vectors were independent and there were no assumed relationship among the site parameters, then again, there would be no additional information in the neighboring sites useful for estimation at site 1. Conversely, potential useful information at neighboring sites exists if there is some dependence among the site vectors and/or some assumed structure among the site parameters. We illustrate with a simple example and see what sort of gain is possible using regional analysis.

Example 3.7–Multivariate Log Normal For our model let us assume that $\underline{X}_{k1},...,\underline{X}_{kn}$ are independent, which assumes independence over years. Further, let us assume stationarity over years so that all $\underline{X}_{kj}$'s have the same distribution and assume that the common distribution is a multivariate two parameter log normal. That is, assume

$$\begin{bmatrix} \ln X_{1j} \\ \vdots \\ \ln X_{kj} \end{bmatrix} \sim \text{MVN} \begin{bmatrix} \mu_1 \\ \vdots \\ \mu_k \end{bmatrix}, \begin{bmatrix} \sigma_1^2 & & \\ & \ddots & \cdots \rho_{ij}\sigma_i\sigma_j \\ & & \sigma_k^2 \end{bmatrix}$$

In general, our parameter θ is $2k + (1/2)(k^2 - k)$ dimensional, there being k μ_i's, k σ_i's, and $(1/2)(k^2 - k)\rho_{ij}$'s. Site i has a marginal log normal distribution with parameters μ_i and σ_i. We are interested in estimating quantiles at site 1, which are given by

$$\xi_q = \exp[\mu_1 + z_q \sigma_1],$$

and are readily obtainable from the assumed log normal model. Now MLEs of all parameters can be found and in particular the MLE of the desired ξ_q can be obtained. It will depend only on site 1 data, and, consequently, *dependence alone* contributes nothing more than would a single site analysis for this model.

Let us assume some structure among the site parameters to examine the effect of such an assumption. As a simple assumed structure among the site parameters, let us assume that all μ_i are proportional to μ and all σ_i are proportional to σ with known proportionality constants. The k μ_i's are reduced to a single parameter μ and the k σ_i's to the single σ. (If one were to try to rationalize such an assumption one could hypothesize that the proportionality constants depend on known basin characteristics.) Let C_i and D_i, $i = 1,...,k$, denote the known proportionality constants for the μ_i's and the σ_i's, respectively; that is

$$\mu_i = C_i \mu \text{ and } \sigma_i = D_i \sigma \text{ for } i = 1,...,k.$$

An estimate of the desired $\xi_q = \exp[\mu_1 + z_q \sigma_1]$ can be obtained if estimates of μ_1 and σ_1 are found. So temporarily, let's concentrate on estimating μ_1. Now

$$\frac{1}{nC_i} \sum_{j=1}^{n} \ln X_{ij}$$

is an unbiased estimator of μ for each i, so

$$\frac{1}{k}\sum_{i=1}^{k}\frac{C_1}{nC_i}\sum_{j=1}^{n}\ell n\, X_{ij}$$

is an unbiased estimator of μ_1 that uses all the regional data:

$$\operatorname{var}\left[\frac{1}{k}\sum_{i=1}^{k}\frac{C_1}{nC_i}\sum_{j=1}^{n}\ell n\, X_{ij}\right]$$

$$= \operatorname{var}\left[\frac{1}{n}\sum_{i=1}^{n}\frac{1}{k}\sum_{j=1}^{k}\frac{C_1}{C_i}\ell n\, X_{ij}\right]$$

$$= \frac{1}{n}\operatorname{var}\left[\frac{1}{k}\sum_{i=1}^{k}\frac{C_1}{C_i}\ell n\, X_{ij}\right]$$

$$= \frac{1}{n}\left\{\frac{1}{k^2}\sum_{r=1}^{k}\sum_{s=1}^{k}\frac{C_1^2}{C_r C_s}\operatorname{cov}[\ell n\, \xi_{rj}, \ell n\, X_{sj}]\right\}$$

(3.34)
$$= \frac{\sigma^2}{nk^2}\sum_{r=1}^{k}\sum_{s=1}^{k}\frac{C_1^2}{C_r C_s}D_r D_s \rho_{r,s}.$$

This variance could now be compared to the single site estimator variance, which is $D_1^2\sigma^2/n$, for various choices of C_j's, D_j's, and $\rho_{i,j}$'s. One can see that the best one can hope to do is to have the multisite variance $1/k$ times the single site variance, which would occur when the double sum over r and s in (3.34) behaves like k. Such gain would be inherited by the estimator of the desired quantile.

The not-surprising end result is that regional analysis can, under an assumption of a strong relationship among site parameters and no cross-correlation, essentially expand the effective data set from that at the single site to the entire region; that is, from n observations to nk observations. The corresponding reduction in standard error of estimate is by a factor of $\sqrt{k}$. In practice the gain due to regionalization will lie somewhere between no gain and the full gain possibly obtainable with no cross-correlation and a strong assumed structure among the site parameters.

A variety of regionalization studies have recently been performed. Lettenmaier and Potter (1985) propose a regional model and report on the relative performance of several methods of regionalization under the assumption that the proposed regional model is correct.

Hosking et al. (1985b) and Wallis and Wood (1985) also report on regional analysis based on index flood methods. Lettenmaier et al. (1986) performed an extensive simulation study designed to explore the robustness of selected regional and single site estimation procedures with respect to: (a) the assumed underlying model; (b) moment (including higher moments) heterogeneity over sites; and (c) variations in record length over sites. Wiltshire (1986a) gives a procedure for classifying basins into distinct, homogeneous groups for regional flood frequency analysis. It is based on coefficients of variation and does not use skewness. Further it does not give unique solutions. Also, Wiltshire (1986b) gives two tests for regional homogeneity and applies them in Wiltshire (1986c).

These regionalization studies have focused increased attention on the idea that hydrologically homogeneous regions need to be delineated in terms of hydrologically meaningful basin characteristics and flood statistics as well as in terms of geographic location. Early regionalization efforts using index flood methods were hampered by difficulties in defining geographic regions in which all sites had similarly shaped frequency curves. These difficulties led to use of regionalization by regression methods, which were better able to represent the relationships between basin characteristics and flood frequency curves (Benson, 1962). These two competing approaches to regionalization can now be reconciled. The regression method describes the set of all flood frequency distributions in a study area in terms of one or more sets of functions of basin parameters. The index-flood approach postulates that the set of all flood-frequency curves can be partitioned into a small number of classes, each one of which can be characterized by a single regional frequency distribution or quantile function. Depending on the number of classes, there will be more or less within-class variability around the regional distributions. For each regional-distribution class, the regional-regression functions define a corresponding set of basin-characteristic values that produce distributions in that class. These sets of basin characteristics thus define hydrologically homogeneous regions: each basin in a region has a frequency distribution belonging to the same regional-distribution class, even though individual parameter values may vary from basin to basin. These regions, moreover, are not simply geographical areas, and it may be difficult to portray them effectively on maps.

Much of the recent activity on regional flood analysis is based on *index flood methods.* In such methods, the data at each site are "normalized" (often using the at-site-mean) and then using all (over all sites) "normalized" data, the parameters of an assumed parametric model for the regional normalized flood are estimated. The actual estimation technique used may depend on the assumed model. Once the regional parameters are estimated, the distribution at each of the sites is assumed to be the same, except for the factor used in the "normalization." Such an assumption is strong and similar to that made in Example 3.7.

The following is an example, more detailed than earlier ones, that considers and illustrates index flood methods for three models.

Example 3.8—Three Flood Models Let N denote the number of sites and let $\xi_i(q)$ be the quantile function of the annual peak flood distribution at site i. This quantile function depends on the flood distribution at site i even though it does not appear in the notation. Let μ_i be the mean at site i. Let $\xi(q)$ be an assumed regional quantile function. Note that the notation here is different from that used earlier where q was subscripted. Here q is used in the argument position of the quantile function to stress that aspect of the function.

The *index flood* assumption is that

$$(3.35) \qquad \xi_i(q) = \mu_i \xi(q).$$

That is, the site-i quantile function is assumed to equal the site-i mean times the regional quantile function. This is a strong assumption and imposes constraints on the site distributions that are mentioned later.

To estimate the site-i quantile function, $\xi_i(q)$, an estimate of μ_i and an estimate of $\xi(q)$ are needed. Estimate μ_i by $\hat{\mu}_i$, the sample mean at site i. To estimate $\xi(q)$, first obtain the sample PWMs at each site and then average (possibly a weighted average) these sample PWMs to get regional sample PWMs. Use these regional sample PWMs to estimate the parameters of an assumed regional distribution with quantile function $\xi(q)$. Replace the parameters in $\xi(q)$ by the estimated parameters to produce $\hat{\xi}(q)$. Finally,

$$(3.36) \qquad \hat{\xi}_i(q) = \hat{\mu}_i \times \hat{\xi}(q).$$

The three index flood models denoted by WAK/R, GEV-1, and GEV-2 can now be defined. WAK/R is the above-described index flood scheme with $\xi(q)$ defined by

(3.37) $$\xi(q) = m + a[1-(1-q)^b] - c[1-(1-q)^{-d}].$$

$\xi(q)$ is the quantile function of the Wakeby distribution, which is a five-parameter distribution defined by its quantile function (which is the inverse of a cdf). An account of the Wakeby distribution and its properties is given in Hosking (1986b).

GEV-1 is the above-described index flood scheme with an assumed GEV for the regional distribution. Here

(3.38) $$\xi(q) = \alpha + (\beta/\kappa)\{1 - [-\ell n\ q]^\kappa\}.$$

[See equation (3.12) and Example 3.3.]

GEV-2 is similar to GEV-1 inasmuch as equation (3.35) is assumed and GEV is the assumed regional distribution. The two models differ in the method used to estimate the parameters. For GEV-2 the location and scale parameters, α_i and β_i in equation 3.12, are estimated for each site using site data and the parameter κ is estimated regionally.

All three procedures are somewhat ad hoc, inasmuch as they possess strong internal constraints under certain assumptions; however, they may provide a useful and robust approach. To illustrate the type of constraint that can arise, suppose that each of the i sites has a GEV distribution with, say, parameters α_i, β_i, and κ_i. Under index flood assumption equation (3.35), one can deduce that $\kappa_i = \kappa_j$ and $\alpha_i/\beta_i = \alpha_j/\beta_j$ for $i \neq j$. $\kappa_i = \kappa_j$ says that all the sites have the same shape parameter, and $\alpha_i/\beta_i = \alpha_j/\beta_j$ says that all sites have the same location to scale ratio, which makes the site distributions quite homogeneous.

A simulation study was performed to illustrate the performance of the three regional procedures. Again the emphasis is on illustration. It is not intended that the numbers that appear in this simulation resemble nature. In fact, it seems unlikely that one would ever find such homogeneity over so many stations with such long records.

Monte Carlo Experiments

Consider five GEV Floodsets (Table 2), for one of which (Floodset 1) a more detailed specification is given in Table 3. Floodsets

TABLE 2 Summary Description of Five Floodsets

Floodset Number	N	n	ρ	heterogeneous
1	41	100	0.00	yes
2	41	100	0.00	no
3	82	100	0.00	no
4	41	200	0.00	no
5	41	100	0.26	yes

1, 2, and 5 have 4,100 station-years of data ($n = 100$ years at $n = 41$ stations). All sites have high coefficient of variation (Cv) and high skew. Floodset 2 is similar to Floodset 1, but each site has a distribution identical to that of site 21 of Floodset 1. Floodset 3 is similar to 2, but with 82 sites, while 4 is similar to 2, but with $n = 200$ for all sites. Floodset 5 is similar to 1, but with an average cross-correlation coefficient among annual floods at each site of .26.

The five Floodsets being considered here have very high Cv's and skewness and are representative of the extreme flood potentials that occur in portions of the arid southwestern United States.

The correlation between sites i and j of Floodset 5 is

$$(3.39) \qquad \rho_{i,j} = \exp(-a \times d_{ij}),$$

where d_{ij} are iid from a uniform distribution, and a is chosen to give $\bar{\rho} = 0.26$. The simulation data were generated as multivariate normal and then transformed to GEV.

To allow for ease of comparison between sites that have different $\hat{\xi}_i(q)$, the $\xi_i(q)$ were scaled as

$$(3.40) \qquad \hat{Z}_i(T) = \frac{\hat{\xi}_i(q) - \xi_i(q)}{\xi_i(q)}.$$

Similarly, to allow for comparisons of the $\xi(q)$ between Floodsets, scaling using equation (3.41) was performed.

$$(3.41) \qquad \hat{z}(T) = \frac{\hat{\xi}(q) - \xi(q)}{\xi(q)}.$$

Here, T is the return period, i.e., $T = 1/p$ and $p = 1 - q$.

TABLE 3 Description of Floodset 1

Site	κ	β	α	Mean	S.D.	C.V.	Skew	Kurtosis	N	Probabilities 0.9900	0.9990	0.9999
											Quantiles	
1	-0.176900	17.510	10.52	24.3	30.37	1.25	3.00	29.4	100.	134.9	247.4	416.4
2	-0.175557	17.524	10.76	24.5	30.31	1.24	2.97	28.7	100	134.8	246.6	413.8
3	-0.174214	17.538	11.01	24.7	30.24	1.22	2.95	28.0	100	134.7	245.7	411.2
4	-0.172871	17.552	11.25	25.0	30.18	1.21	2.92	27.4	100	134.6	244.8	408.7
5	-0.171528	17.567	11.50	25.2	30.12	1.20	2.90	26.7	100	134.5	244.0	406.2
6	-0.170185	17.581	11.74	25.4	30.05	1.18	2.87	26.1	100	134.4	243.1	403.7
7	-0.168842	17.595	11.99	25.6	29.99	1.17	2.85	25.6	100	134.4	242.3	401.3
8	-0.167498	17.609	12.23	25.9	29.93	1.16	2.82	25.0	100	134.3	241.4	398.8
9	-0.166155	17.623	12.48	26.1	29.87	1.15	2.80	24.5	100	134.2	240.6	396.4
10	-0.164812	17.637	12.72	26.3	29.81	1.13	2.77	24.0	100	134.1	239.8	394.0
11	-0.163469	17.651	12.97	26.5	29.74	1.12	2.75	23.5	100	134.0	239.0	391.6
12	-0.162126	17.665	13.21	26.7	29.68	1.11	2.73	23.0	100	134.0	238.1	389.3
13	-0.160783	17.680	13.45	27.0	29.62	1.10	2.70	22.5	100	133.9	237.3	387.0
14	-0.159440	17.694	13.70	27.2	29.57	1.09	2.68	22.1	100	133.8	236.5	384.6
15	-0.158097	17.708	13.94	27.4	29.51	1.08	2.66	21.6	100	133.7	235.7	382.4
16	-0.156754	17.722	14.19	27.6	29.45	1.07	2.64	21.2	100	133.7	235.0	380.1
17	-0.155411	17.736	14.43	27.9	29.39	1.05	2.61	20.8	100	133.6	234.2	377.9
18	-0.154068	17.750	14.68	28.1	29.33	1.04	2.59	20.4	100	133.5	233.4	375.6
19	-0.152725	17.764	14.92	28.3	29.28	1.03	2.57	20.0	100	133.4	232.6	373.4
20	-0.151382	17.778	15.17	28.5	29.22	1.02	2.55	19.6	100	133.4	231.9	371.2
21	-0.150038	17.793	15.41	28.8	29.16	1.01	2.53	19.3	100	133.3	231.1	369.1

22	-0.148695	17.807	15.65	29.0	29.11	1.00	2.51	18.9	100	133.2	230.4	366.9
23	-0.147352	17.821	15.90	29.2	29.05	0.99	2.49	18.6	100	133.2	229.6	364.8
24	-0.146009	17.835	16.14	29.4	29.00	0.99	2.47	18.3	100	133.1	228.9	362.7
25	-0.144666	17.849	16.39	29.6	28.94	0.98	2.45	17.9	100	133.0	228.1	360.6
26	-0.143323	17.863	16.63	29.9	28.89	0.97	2.43	17.6	100	133.0	227.4	358.6
27	-0.141980	17.877	16.88	30.1	28.83	0.96	2.41	17.3	100	132.9	226.7	356.5
28	-0.140637	17.891	17.12	30.3	28.78	0.95	2.39	17.0	100	132.9	226.0	354.5
29	-0.139294	17.905	17.37	30.5	28.73	0.94	2.38	16.8	100	132.8	225.3	352.5
30	-0.137951	17.920	17.61	30.8	28.67	0.93	2.36	16.5	100	132.7	224.6	350.5
31	-0.136608	17.934	17.85	31.0	28.62	0.92	2.34	16.2	100	132.7	223.9	348.6
32	-0.135265	17.948	18.10	31.2	28.57	0.92	2.32	16.0	100	132.6	223.2	346.6
33	-0.133922	17.962	18.34	31.4	28.52	0.91	2.30	15.7	100	132.6	222.5	344.7
34	-0.132579	17.976	18.59	31.7	28.47	0.90	2.29	15.5	100	132.5	221.8	342.8
35	-0.131235	17.990	18.83	31.9	28.42	0.89	2.27	15.2	100	132.5	221.1	340.9
36	-0.129892	18.004	19.08	32.1	28.37	0.88	2.25	15.0	100	132.4	220.4	339.0
37	-0.128549	19.018	19.32	32.3	28.32	0.88	2.23	14.7	100	132.4	219.8	337.1
38	-0.127206	18.033	19.57	32.6	28.27	0.87	2.22	14.5	100	132.3	219.1	335.3
39	-0.125863	18.047	19.81	32.8	28.22	0.86	2.20	14.3	100	132.3	218.5	333.5
40	-0.124520	18.061	20.06	33.0	28.17	0.85	2.18	14.1	100	132.2	217.8	331.7
41	-0.123177	18.075	20.30	33.2	28.12	0.85	2.17	13.9	100	132.2	217.2	329.9

Total Station-years = 4100

SCALED REGIONAL CURVE (HARMONIC MEAN)   4.500   7.738   12.266

Floodset 1 and 2: $\hat{Z}_i(T)$

For comparison purposes, it may be helpful to consider $\hat{Z}_{21}$ of Floodset 2 fitted by an at-site method (unbiased PWM). As shown in Figure 3, the estimated quantiles from successive samples are highly variable, with 25 percent of the $T_{10,000}$ estimates being more than 27 percent too big, and a further quarter being below 30 percent of their true value. This confirms what was stated earlier. For large T estimates are highly variable.

For Floodset 2 (using all sites), the GEV-1 and GEV-2 results for site 21 are given in Figures 4 and 5, respectively, where it can be seen that the estimators are unbiased and remarkably stable. Figure 6 shows the comparable analysis to that of Figures 3-5 but for WAK/R. The WAK/R results exhibit little variability between successive estimates but the bias of −15 percent for the $T_{10,000}$ event may be unacceptably large for the purpose of estimating the recurrence interval of a probable maximum flood (PMF). Floodset 2 is homogeneous, and as such the results obtained for each site are, except for sampling variability, identical, but this situation does not extend to Floodset 1 and 5. The quantile estimates for nonrepresentative sites in an heterogeneous Floodset may be severely biased, but what is rather surprising and very useful in the context of this report is the phenomenon that the quantile estimates for representative sites in a heterogeneous floodset are essentially as good as those obtainable from homogeneous Floodsets (for instance, compare Figure 7 with Figure 4, or Figure 8 with Figure 5).

With GEV-1 and WAK/R analyses of heterogeneous Floodsets some sites will have negatively biased $\hat{\xi}(q)$ [e.g., Figure 9 gives $\hat{Z}_1(T)$], while others will be positively biased [Figure 10 shows $\hat{Z}_{41}(T)$ for Floodset 1 and GEV-1].

If the Floodset is large enough and the $\overline{Cv}$ is high, better results for the less representative sites (if needed) may be obtained by GEV-2 (compare Figure 11 and 9, and 12 with 10). It is clear that heterogeneity in a Floodset need not cause undue alarm, providing that the design site is more or less representative of the set as a whole. Finally, let it be recognized that one need not necessarily abandon an index flood approach to flood frequency analysis just because there is heterogeneity among the floodset catchments.

Floodset 5: $\hat{Z}_i(F)$

As suggested earlier the observed correlation among the sites

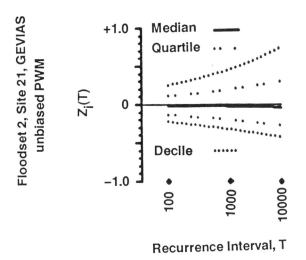

FIGURE 3 At-site GEV quantile estimates (based upon unbiased PWM's) for site 21 of Floodset 2, showing the median as well as the upper and lower quartile and decile values.

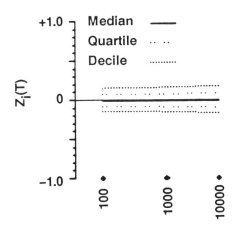

FIGURE 4 GEV-1 quantile estimates (based upon unbiased PWM's) for site 21 of Floodset 2, showing the median as well as the upper and lower quartile and decile values.

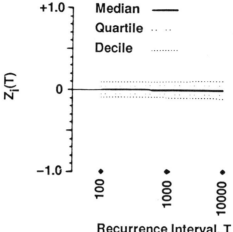

FIGURE 5 GEV-2 quantile estimates (based upon unbiased PWM's) for site 21 of Floodset 2, showing the median as well as the upper and lower quartile and decile values.

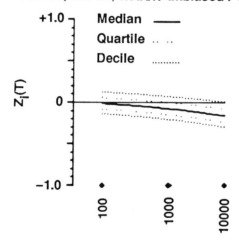

FIGURE 6 WAK/R quantile estimates (based upon unbiased PWM's) for site 21 of Floodset 2, showing the median as well as the upper and lower quartile and decile values.

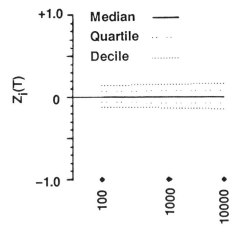

FIGURE 7 GEV-1 quantile estimates (based upon unbiased PWM's) for site 21 of Floodset 1, showing the median as well as the upper and lower quartile and decile values.

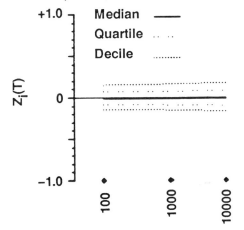

FIGURE 8 GEV-2 quantile estimates (based upon unbiased PWM's) for site 21 of Floodset 1, showing the median as well as the upper and lower quartile and decile values.

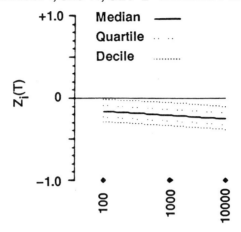

FIGURE 9 GEV-1 quantile estimates (based upon unbiased PWM's) for site 1 of Floodset 1, showing the median as well as the upper and lower quartile and decile values.

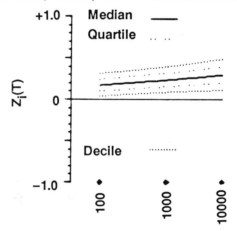

FIGURE 10 GEV-1 quantile estimates (based upon unbiased PWM's) for site 41 of Floodset 1, showing the median as well as the upper and lower quartile and decile values.

Floodset 1, Site 21, GEV-2 unbiased PWM

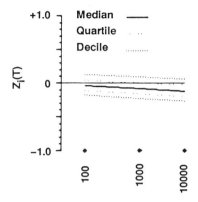

FIGURE 11 GEV-2 quantile estimates (based upon unbiased PWM's) for site 1 of Floodset 1, showing the median as well as the upper and lower quartile and decile values.

Floodset 1, Site 1, GEV-1 unbiased PWM

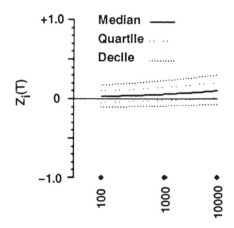

FIGURE 12 GEV-2 quantile estimates (based upon unbiased PWM's) for site 41 of Floodset 1, showing the median as well as the upper and lower quartile and decile values.

Floodset 1, Site 41, GEV-1 unbiased PWM

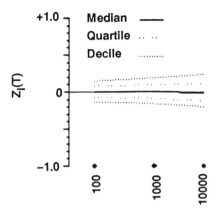

FIGURE 13 GEV-1 quantile estimates (based upon unbiased PWM's) for site 21 of Floodset 5, showing the median as well as the upper and lower quartile and decile values.

of real world Floodsets is usually low, and as such need not be of great concern to those making the types of analyses presented here. While a $\bar{\rho}$ of 0.26 does cause a deterioration in the $\hat{\xi}_i(q)$, it is not of the magnitude that is implied by Equation (3.27). To illustrate the actual degradation likely to be experienced as a result of correlation compare Figures 7 and 13.

Floodset 3 and 4: $\hat{Z}_i(T)$ and $\hat{z}(T)$

It has been stated that if one had 40 years of record at 100 gages in a region, then one would have 4,000 station-years of data. However, this is still far less than the 10,000 to 1,000,000 station-years of data that would be desirable to reliably and credibly estimate floods with return periods on the order of 10^4–10^6 years (National Research Council, 1985). Morever, when robust regional estimators are used the reduction in the size of the confidence invervals may be rather small after the data sets grows larger than a few thousand station-years.

As Floodsets increase in size the $\hat{\xi}(q)$ rapidly approach an estimate that changes only incrementally as further sites are added. For

Floodset 1 and GEV-1 the median and upper and lower deciles of the growth curve, $\hat{z}(T)$ are shown on Figure 14. It is also interesting to compare Figure 14 and Figure 15, which shows similar results from the addition of a further 41 sites, each with an n of 100. It is clear that once the N of a Floodset gets large the addition of more sites will increase the accuracy of $\hat{\xi}_i(q)$ only very slowly.

It would appear that if we wish to increase the accuracy of $\hat{\xi}_i(q)$ we must devise a method for improving the estimate of the at-site component of regionalization. Regressions or similar methods from the ungaged watershed approach are possibilities, and longer records might be another way to improve $\hat{\xi}_i(q)$. As can be seen from comparing Figure 16 and Figure 7 the improvements in $\hat{\xi}_i$ that result from longer records may also be slow in arriving. Even if very long records exist, one should be wary of nonstationarities.

It would appear that the chief benefits of accruing an extremely large floodset would be, first, to allow for a more judicious selection of floodset sites, and second, to serve as a data base for better estimation of the at-site component of the analysis.

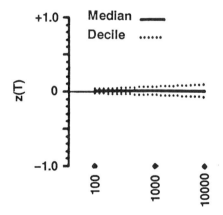

FIGURE 14 GEV-1 growth curve estimates (based upon unbiased PWM's), for Floodset 2, showing the median as well as the upper and lower decile values.

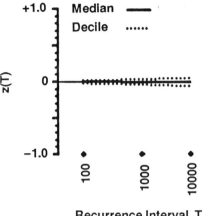

FIGURE 15 GEV-1 growth curve estimates (based upon unbiased PWM's), for Floodset 3, showing the median as well as the upper and lower decile values.

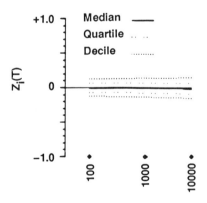

FIGURE 16 GEV-1 quantile estimates (based upon unbiased PWM's) for site 21 of Floodset 4, showing the median as well as the upper and lower quartile and decile values.

ASSESSMENT

In a regional analysis, a single storm/hydrologic event may produce annual peak readings at all or most sites in a region. For such a year one might expect the readings to be highly correlated, and, consequently, the regional information is not much more than single site information. Such dependence detracts from any gain that might be obtained in assuming a strong structure among site parameters. For instance, if the $\rho_{r,s}$ are large then the double sum in (3.34) may grow like k^2 and negate the k^2 in the denominator. One would not be much better off than by doing a single-site analysis.

Another method that has potential for use in a regional analysis is that of employing empirical Bayesian techniques. In a Bayesian analysis some prior distribution on the parameter space is assumed. Regionalization might suggest that the various sites have the same prior distribution. The neighboring sites could then be used to estimate the prior distribution, and then a Bayes estimator of the parameters at the site of interest could be obtained based on the estimated prior distribution, using empirical Bayes methods.

Just as historical/paleoflood data ought not to be ignored, regional data should not be either. The problem in regional analysis is twofold. First, how does one select/identify and check appropriate joint parametric models? One has difficulty in selecting a single site model, and selecting a joint (versus marginal) model is an order of magnitude more difficult. Second, since the potential gain in multi-site analysis is primarily due to the assumed structure among the site parameters, how does one select and verify such a structure?

A STRATEGY FOR ESTIMATING $\xi_i(q)$

From the preceding discussions it would appear that a strategy for estimating $\xi_i(q)$ for a specific site i of interest would be to:

1. Compile a large set of flood records from sites that are hydrologically similar to the site of interest. (See chapter 5 discussion on streamflow data sample size.)

2. Depending upon the amount and kind of data, pick a robust regionalization procedure including the model and estimation method.

3. Estimate $\xi_i(q)$ and obtain the standard error of estimator via Monte Carlo, jackknife, or bootstrap type simulations (Efron, 1982), as appropriate.

4. Check to see that the estimation method does not have excessive bias at the desired return period and check consistency.

5. Using alternative parent distributions (depending upon how sure one is about the choice in step 2, check to see that the method selected is sufficiently robust for the quantile or quantiles of interest.

CONCLUSION

As noted in the beginning of this chapter, the essential ingredients of *regionalization, historical/paleoflood data,* and *robustness* associated with the estimation problem have been discussed more or less individually. These should now be combined. With few exceptions the discussion has focused on the annual peak data case. Of course, one has more than just annual peak data and the utility of the other data, such as flood volumes and durations, should be explored. Most of the statistical considerations discussed in this chapter apply also to these other types of flood data.

The bulk of the cited literature is very recent and does not contain a satisfactory general solution. If one were willing to accept a specific parametric model (capable of handling regionalization and incorporating historical/paleoflood data), then the tools are in place to derive the desired estimate and obtain a measure of the precision of the estimator. Since there is no way in practice that one is going to be able to check the assumed model, one has to adopt a procedure that is robust. The work on tail behavior is promising and needs to be articulated in a regional setting; it is ready built to handle historical data that can be interpreted as censored random samples. Parametric modeling and analysis still have a role, and can be used as a base on which to compare robust procedures.

4

Runoff Modeling Methods

INTRODUCTION

This chapter considers methods for estimating flood probabilities using runoff models. The basic idea is as follows. A runoff model is selected and calibrated for the catchment in question. Then meteorologic inputs are developed, the most important of which is rainfall. Associated with these inputs is some sort of probabilistic structure, either implicit or explicit. The runoff model is run with the meteorologic inputs, and streamflows are obtained. Based on these streamflows, flood exceedance probabilities are estimated, with the method of estimation depending on the assumed probabilistic structure of the meteorologic inputs, especially rainfall.

The most obvious advantage of a runoff modeling method is that it can be used at locations at which there is little or no streamflow data (though the calibration of the runoff model depends on the availability of some data). For this reason runoff modeling methods for estimating flood probabilities are in common use, though nearly all applications involve relatively common floods (with exceedance probabilities that are greater than or equal to 10^{-2}). A second advantage of runoff models is their ability to generate entire hydrographs, which may be critical in applications for which flood volumes are important. Finally, with runoff models it is possible to account explicitly for changes in land use.

For our problem—the estimation of exceedance probabilities of extremely rare floods—the greatest potential advantage of runoff modeling has to do with regionalization. In a region of homogeneous meteorology, the differences in population flood frequency distributions at various sites will be due largely to differences in the physical characteristics of the corresponding catchments, rather than to differences in meteorology. Hence it should be advantageous to regionalize the meteorologic factors, especially rainfall, and then use a runoff model to account for drainage basin characteristics.

Since their development, runoff models have been used in a variety of ways to estimate flood probabilities. The applications generally have been limited to floods with exceedance probabilities greater than or equal to 10^{-2} or to computations of probable maximum floods.

The use of a runoff model to estimate flood probabilities involves several distinct steps.

- First, a runoff model must be selected and calibrated for the catchment of interest. Either a continuous model or an event model can be used. A continuous runoff model simulates both storm events and interstorm periods. Hence it continuously models all forms of moisture storage, such as soil moisture, groundwater storage, and snowpack. An event model simulates only storm events.

- Once a runoff model has been calibrated, meteorologic inputs and an associated probabilistic structure must be developed. A continuous runoff model requires a continuous record of rainfall, as well as information on conditions that affect the depletion of moisture storage, such as potential evapotranspiration, temperature, and solar radiation. An event model requires rainfall amounts only for the storm events of interest.

- Before the runoff model can be run using the meteorologic inputs, initial conditions must be specified for various model states, such as moisture storage. This so-called specification of antecedent conditions is not critical in the case of a continuous runoff model, since the impact of antecedent conditions usually becomes negligible after the model has simulated a few weeks or months of streamflow. In the case of an event model, antecedent conditions must be specified for each storm event that is simulated. Hence their specification is a critical part of the process.

- Next the runoff model is run using the meteorologic inputs and the specified antecedent conditions.

- This is followed by estimation of the exceedance probabilities of the resulting peak flood discharges. The procedure for doing this depends on the nature of the assumed probabilistic structure of the rainfall inputs. In the case of an event model, this step must also account for the specification of antecedent conditions.

- The final step is to estimate the uncertainty in the computed probability distribution.

Many strategies have been proposed and used to conduct each of the previously described steps. No attempt is made to review the literature on this subject. Instead we focus on those strategies which are most appropriate for estimating exceedance probabilities of extreme floods. In some cases we also propose new strategies that we believe to be promising. We begin by considering runoff modeling. Then we focus on the development of meteorologic inputs, particularly rainfall. The critical issue here is how to regionalize the probabilistic structure that is developed for the rainfall input. We follow with a discussion of the estimation of probabilities of the simulated peak flood discharges, and conclude with the problem of evaluating the uncertainty in the resulting estimated exceedance probabilities.

RUNOFF MODELS

Runoff models (also known as rainfall-runoff models) are hydrologic models which simulate streamflow at one or more locations in a catchment. Runoff models require three types of input: meteorologic data, antecedent conditions, and catchment characteristics. Hourly rainfall during a storm is an example of meteorologic data. Soil moisture at the start of the storm is an antecedent condition. Stream channel length and slope are typical catchment characteristics.

A runoff model is a collection of algorithms that mimic the hydrologic processes involved in the transformation of rainfall into streamflow. The most important of these processes are the infiltration of water into the soil during storm events and the movement or "routing" of water over the land, through the subsurface, and through the channel network. The algorithms composing a runoff model contain parameters which represent catchment characteristics. Many of these parameters represent physical characteristics that can be directly measured, such as channel characteristics. Other parameters, such as those governing soil infiltration, are usually specified to produce the best possible match between simulated and observed streamflow (where available). This process is called calibration. When calibrating models for use in estimating extreme floods it is important to calibrate against the largest floods of record so as to minimize the degree of extrapolation.

In addition to simulating hydrologic processes during storm events, a continuous model also simulates processes during interstorm events. The most important of these interstorm processes are depletion of soil moisture by evapotranspiration and drainage and the discharge of ground water into the stream channel (baseflow). When a continuous rainfall record is used as input to a continuous runoff model, antecedent conditions, such as initial soil moisture, need only be specified at the start of the simulation. The model will continuously account for changes in soil moisture through the simulation. By starting a continuous simulation well before the time of occurrence of the first storm of interest it is possible to minimize the impact of the choice of antecedent conditions. An event model, on the other hand, simulates only storm events (or sequences of storm events). Hence the specification of antecedent conditions is of much greater importance.

DEVELOPMENT OF METEOROLOGIC INPUTS

Once a runoff model has been selected and calibrated, the next step in using the model to estimate flood probabilities is to develop meteorologic inputs and some sort of associated probabilistic structure. In the case of an event model applied in a region in which floods are due to rainfall alone, only rainfall inputs are needed. A continuous model requires additional inputs to account for the evapotranspiration process. For applications in regions where snowmelt is a factor, additional meteorologic inputs are required to model that process. In this section we focus on the development of rainfall inputs. At the end of this section a few comments are provided on other meteorologic inputs.

Rainfall Inputs

Several strategies have been used to develop rainfall inputs for use in estimating flood probabilities by runoff modeling. In only a minority of cases has the focus been on extreme floods. Hence we will focus more on what could be done rather than what has been done. We have classified potential methods into three groups, depending on whether they are based on: (1) direct use of actual data, (2) stochastic rainfall models, or (3) synthetic storms. This classification is motivated by the probability structure associated with each kind of input. When actual data are used as input to a runoff model, it is assumed that the data represent a random sample of possible rainfall. Stochastic rainfall models are developed to reproduce explicitly an inferred probabilistic structure of rainfall. The use of synthetic storms is based on the concept of a storm event for which explicit exceedance probabilities can be estimated. Note that this classification of potential methods is somewhat arbitrary. It is not difficult to conceive of methods that have characteristics of more than one class.

It should be kept in mind that rainfall varies over both space and time in an extremely complex manner, which to date is only partly understood. Furthermore, over certain spatial and temporal scales this variability translates into comparable variability in the runoff process. The strategies discussed below for developing rainfall inputs differ significantly in the extent to which they account for this temporal and spatial variability.

Direct Use of Actual Data

The most obvious approach to using runoff modeling to estimate flood probabilities is to run a calibrated continuous runoff model using one or more long continuous records of rainfall that have been measured in or near the catchment of interest. It is assumed that the observed rainfall records can be considered to be a random sample from a population of rainfall sequences. A modification of this approach is to restrict the modeling to storm events that are judged to be significant. For example, some simple criteria can be used to select the most significant storms from each year of the observational record. These storms can then be run through a calibrated event runoff model, and the largest peak discharge in each year retained for subsequent analysis. It is assumed that the selection criteria ensure that the storm that would produce the annual peak would be included in the selected set. This approach had been demonstrated as early as 1957 (Paulhus and Miller, 1957) and has been used routinely by the U.S. Geological Survey (e.g., Hauth, 1974; Thomas, 1982, 1986a,b; Krug and Goddard, 1985). In these applications there has been no attempt at regionalization; unless this can be done, the direct use of station data is not a promising method for estimating exceedance probabilities of extreme floods, in view of the relatively short length of available rainfall records.

Stochastic Models

Stochastic models have several conceptual advantages for use in generating rainfall inputs to runoff models. Generic model structures can be developed based on data from long records or dense networks. This in itself is a form of regionalization. The resulting model can then be calibrated for the site of interest, allowing an additional opportunity for regionalization.

Stochastic rainfall models fall into three broad groups. One-dimensional temporal models simulate rainfall at a point. Multidimensional space-time models simulate rainfall in both space and time, commonly employing random field theory to preserve the known regularities in the space-time structure of rainfall. Intermediate between these groups are multistation models; these are based on temporal models that are fitted to station data and linked together to provide some degree of spatial structure by preserving certain cross-correlational properties.

Of these three groups, one-dimensional temporal models have received the most attention. [See Waymire and Gupta (1981) for a recent survey.] Due to intermittency of the rainfall process at small time scales, such as hours or days, the general approach is to model separately the occurrence of rainfall events (suitably defined), their duration, and the amount of rainfall in each event. Models have been constructed for both continuous and discrete time, with the day being the most common interval of discrete models.

Despite the significant body of literature on one-dimensional temporal models, there is some question as to whether they adequately model extreme rainfalls. Valdes, Rodriguez-Iturbe, and Gupta (1985) addressed this question in the following way. Using the multidimensional space-time model of Waymire, Gupta, and Rodriguez-Iturbe (1984), they simulated rainfall traces at a fixed location. They then aggregated the data over various time intervals, fitted the parameters of three well-known one-dimensional temporal models to the aggregated data, and generated rainfall traces from the fitted models. Extreme value analysis of all rainfall traces indicated that none of the temporal models was able to reproduce correctly the upper tail distribution of the multidimensional model on which it was based.

Multidimensional space-time models have been developed in recent years to preserve known regularities in the structure of rainfall in space and time. For example, the previously mentioned model of Waymire, Gupta, and Rodriguez-Iturbe (1984) captures the observed space-time structure of rainfall in extratropical cyclonic storms and is consistent with the empirical results of Zawadzki (1973) concerning the covariance structure of rainfall intensities. Space-time modeling, however, is in its infancy; fully operational models may not be available for several years. Of particular concern is the problem of estimating the large number of parameters that space-time modeling requires. In fact no space-time models have been fitted to actual data. Furthermore, in many practical applications the lack of a sufficiently dense gage network may make space-time modeling infeasible. An alternative is to use a space-time model based on radar data, such as the one developed by Kavvas and Herd (1985). In addition to the issue of parameter estimation, there are two other questions regarding the use of space-time models to estimate the probabilities of rare floods. First, how well do space-time models represent extreme rainfalls? Second, what level of space-time detail is required for runoff modeling? To date, there appears to have been no research on these questions.

Multistation models fall between one-dimensional temporal models and multidimensional space-time models. Generally, they are based on temporal models that have been modified to account for spatial structure. In several cases multi-station models have been developed specifically to provide rainfall inputs to runoff models. Two of these cases are discussed below.

Franz, Kraeger, and Linsley (1986) introduced a multistation rainfall model that they used in a continuous runoff model to estimate probabilities of flood events with recurrence intervals of up to 10^4 years. The rainfall model generates hourly rainfall at two stations. At each station the model is based on a nonobservable background process that is assumed to be Gaussian and autoregressive. Values of the background process yield zero rainfall when they fall below a specified threshold (appropriately defined to preserve the run-properties of the data); values above the threshold are converted into rainfall by a two-parameter nonlinear transformation. The at-site parameters are seasonally estimated to preserve the following characteristics of station rainfall: probability of a storm; probability of rain in any given hour; mean, variance, and distribution of hourly intensity during storms; distribution of storm lengths; serial autocorrelation. The single-station model is generalized to a two-station model through the background process, which is assumed to be bivariate normal. The generalized model is calibrated with station data to preserve the probability of joint occurrences at lag-zero through lag-two. With respect to the specific station characteristics that were targeted for preservation, the performance of the model was deemed satisfactory, except in the case of the distribution of storm lengths. The model also did an adequate job of mimicking the probability distribution of the annual maximum depths for durations of 1, 12, 24, and 36 hours. However, the model was unable to preserve adequately the distribution of interstorm lengths.

Bras et al. (1985) also developed a multistation rainfall model that was used to generate rainfall inputs to an event runoff model. The rainfall model generates the starting time, total depth, and duration of the storm; determines the subbasin in which the storm center is located and the number of subbasins affected by the storm; and partitions the storm depth at 50 percent and 100 percent of the storm duration into corresponding subbasin depths, to preserve the historically observed spatial covariances. (In the example of Bras et al., 17 subbasins were used.) At each subbasin the temporal distribution of the storm (called "storm interior") is generated based on techniques first suggested by Wilkinson and Tavares (1972). All model parameters are estimated seasonally, based on average subbasin rainfall computed from gage data.

In summary, stochastic models offer several conceptual advantages for use in developing rainfall inputs to runoff models to estimate flood probabilities. They can exploit many kinds of data and can incorporate details of the space-time characteristics of rainfall. Regionalization can be employed in both model development and calibration. However, further research is needed to evaluate the ability of stochastic models to represent extreme events.

Synthetic Storms

In both of the previous methods, exceedance probabilities associated with the rainfall inputs are implicit. An observed rainfall record is assumed to represent a random sample. Stochastic models are used to generate random samples. In the case of synthetic storms, exceedance probabilities are explicitly associated with specific storm events.

As previously noted, rainfall varies over space and time in an extremely complex manner. In order to assign exceedance probabilities to specific storm events, it is necessary to simplify observed storms by averaging over space and time. Typically this is done in the following way. First, a storm is defined to have a fixed duration. Sometimes this duration is chosen to equal the characteristic response time of the catchment, though often other times are used. Next, for any given average depth over this duration the exceedance probability is estimated from existing rainfall data using methods discussed below. Finally, a temporal structure is specified for the storm, usually based on observed characteristic temporal distributions of intensity.

The critical step in the development of synthetic storms is the estimation of exceedance probabilities. By far the most commonly used approach for doing this, particularly for small storms (100 mi^2 or less) of exceedance probability 10^{-2} and higher, is based on estimating depth–duration–frequency relationships from station data. We refer to this as the depth–duration–frequency relationship method. A second approach, which has been suggested for use with very low probability storms, is based on a historic storm catalog, such as the one compiled by the National Weather Service and the U.S. Army Corps of Engineers. Methods based on this approach will be referred to as storm transposition methods.

Depth–Duration–Frequency Relationship Method A depth–duration–frequency relationship gives, as a function of duration, the total depth of rainfall at a point that will be exceeded in any year with specified probability. An individual relationship is estimated from data from a single station. The National Weather Service has done this for station data throughout the United States (Hershfield, 1961; Miller et al., 1973) and has interpolated the resulting relationships for use at any location. [See Wenzel (1982) for a description of the procedures which have been used.]

A synthetic storm of specified exceedance probability is constructed in the following way. The depth is obtained from the depth–duration–frequency relationship based on the specified probability and duration. Next, to account for the fact that depth–duration–frequency relationships are based on point data, an area adjustment factor is applied to the rainfall depth. [Factors such as those developed by the Weather Bureau (1958) are commonly used (for areas less than 400 mi^2).] Finally, a method is used to distribute the rainfall depth over time. [Wenzel (1982) discusses several such methods.]

Depth–duration–frequency relationships can be and have been regionalized. The National Weather Service has regionalized depth–duration–frequency relationships for rainfall depths with probabilities down to 10^{-2}. One approach for more extreme rainfall depths is to use one of the regionalization techniques discussed in the previous chapter. Hosking and Wallis (1987) have used such an approach with California rainfall maxima. Richards and Wescott (1986) suggest several approaches for regionalization based on a hybrid series composed of the largest annual point value of rainfall for a specified duration from all stations within a 50-mile radius of the site of interest.

Storm Transposition Methods Storm transposition is a deterministic concept that has been used extensively in the development of estimates of probable maximum precipitation (PMP). Its use here is a natural extension of that application.

In the PMP application, storm transposition is based on the assumption that there exist meteorologically homogeneous regions such that a major storm occurring somewhere in the region could occur anywhere else in the region, with the provision that there may be differences in the averaged depth of rainfall produced based upon differences in moisture potential. In practice, scaling is based on the ratio of moisture potential at the transposed location to that at the observed storm location. Computation of this ratio is commonly performed using precipitable water amounts corresponding to observed maximum persisting 12-hour 1,000-mb dew points at the two sites. Analyses of the latter are found in the *Climatic Atlas of the United States* (Environmental Data Service, 1968), and tables of precipitable water equivalents can be found in the *Manual for Estimation of Probable Maximum Precipitation* (World Meteorological Organization, 1986). In some locations orographic adjustments are also made. For our application, the concept of storm transposition is extended to incorporate the probability of occurrence. Thus, it is assumed that meteorologically homogeneous regions exist such that a major storm of a given magnitude occurring somewhere in the region has the same probability of occurring anywhere else in the region, subject to an appropriate adjustment of the magnitude. Hence, by using extreme storm events that have occurred historically in a meteorologically homogeneous region that is large compared to the catchment of interest, it is possible to "effectively" extend the data base used to estimate storm exceedance probabilities for the catchment.

Ideally, the use of storm transposition to estimate exceedance probabilities of very extreme events would be based on a long-term, complete record of the space-time characteristics of major storms. For the contiguous United States, data have been collected and archived for many of the major rainfall events that have occurred during the last 100 years or more. A similar data base exists for Canada. The U.S. Army Corps of Engineers has developed depth-area–duration tables for about 570 of the United States storms, covering a period between 1875 and 1972. As seen from Figure 17, these storms are unevenly distributed throughout the United States. Limited information is available on an additional 290 storms, covering a period between 1819 and 1984. The location of these storms is shown on Figure 18.

There have been only a few applications of storm transposition to the problem of estimating exceedance probabilities of rare rainfall events. Alexander (1963, 1969) introduced the concept, but provided no examples of its application. Newton and Cripe (1973) applied a version of Alexander's method to watersheds in the Tennessee Valley region; subsequently the basic concept has been used to evaluate flood risks to dams and nuclear power plants in the region. Yankee Atomic Electric Company (1984) presents an original method for using storm transposition to estimate exceedance probabilities of rare rainfall events, and used the method to evaluate the probability of overtopping of a dam on a small catchment in Vermont. In the next section we formalize the concept of storm transposition and outline a general approach for using it in conjunction with a complete historic storm catalog to estimate storm probabilities. We then show how this general approach relates to the methods proposed by Alexander (1963, 1969) and Yankee Atomic Electric Company (1984).

Formulation of a Storm Transposition Model

Consider a catchment of area A_c located in a larger "meteorologically homogeneous" region of area A_h. (In the context of the storm transposition model developed here, the term "meteorologically homogeneous" is defined below. Note that this context differs from that of PMP.) Assume that the critical duration of the catchment is D_c.

Define a "significant storm" to be a rainfall event of duration D_c that is sufficiently large to produce an average rainfall over the

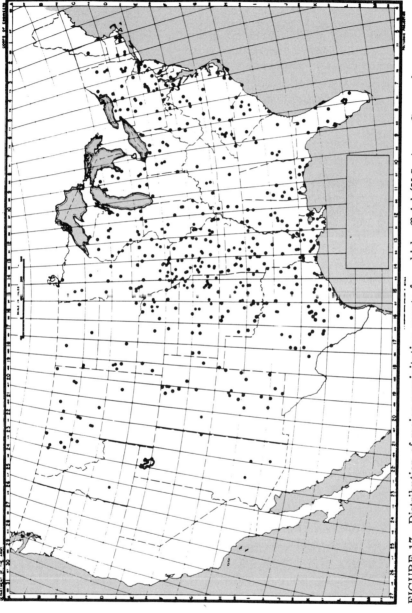

FIGURE 17 Distribution of major precipitation events for which official (U.S. Army Corps of Engineers) depth–area–duration data have been published.

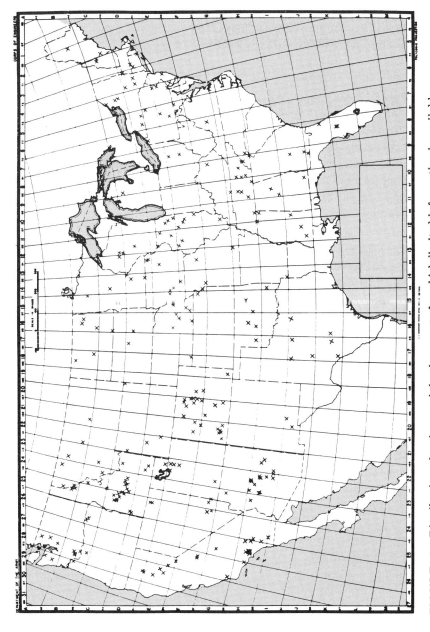

FIGURE 18 Distribution of major precipitation events for which limited information is available.

catchment, if optimally centered, which equals or exceeds a critical value x_c. Assume that x_c is chosen sufficiently large so that the probability of two or more storms occurring in the region in a given year is much smaller than the probability of occurrence of a single storm.

Let X be a random variable representing the largest value of average rainfall occurring on the catchment over an interval D_c in any year. The probability distribution of X is the key result of this analysis; it is subsequently used to estimate exceedance probabilities of extreme annual floods. Note that the random variable depends on both meteorologic factors and the geometry of the catchment. Further, by defining a storm to have a duration D_c we are implicitly assuming that an event of duration D_c "causes" the annual flood in each year in which the threshold x_c is exceeded.

We want to estimate $p(x)$, where

$$p(x) = P\{X \geq x\}.$$

This can be done by conditioning on the occurrence and location of significant storms in the region.

Let (y,z) be cartesian coordinates of an arbitrary point in the homogeneous region. Assume that, given a significant storm has occurred in the homogeneous region, the probability that its center of mass falls in any small region of area $(\Delta y)(\Delta z)$ centered at (y,z) is $(\Delta y)(\Delta z)/A_h$. (This is the first of two assumptions that can be used to define a meteorologically homogeneous region.) Let p_h be the probability that a "significant" storm occurs in the homogeneous region in a given year. Let $p[x,(y,z)]$ be the probability that, given a "significant" storm is centered at (y,z), it will produce an average depth of rainfall on the catchment equalling or exceeding x. Then

(4.1) $$p(x) \approx (p_h/A_h) \int\int_{A_h} p[x,(y,z)]dydz.$$

This relationship is approximate because it ignores the possibility that more than one significant storm will occur in a given year.

The critical assumption used to derive (4.1) is that storm locations have a uniform distribution over A_h. This assumption could be used to define a "meteorologically homogeneous" region. Strict meteorologic homogeneity could also be assumed to imply that the probability $p[x,(y,z)]$ in (4.1) depends on the relative locations of the storm and catchment centers, rather than on the absolute locations. This assumption is not absolutely necessary, provided the dependence of $p[x,(y,z)]$ on absolute catchment location can be specified. This dependence could account for known variations in storm characteristics with location.

Estimation

Assume that over some time period of length N years we have data on *all* significant storms in a meteorologically homogeneous region containing the catchment of interest. (At present the existence of a complete data base for any region is uncertain.) Let m be the number of significant storms. We can estimate p_h by

$$\hat{p}_h = m/N.$$

The probability $p[x,(y,z)]$ can be estimated as follows. Create a square grid on the homogeneous region. Let ΔA be the area of each square and let i be the index of the coordinates of the center of each square (grid centers). Locate the center of each of the m significant storms at each grid center and determine whether or not the storm produces an average depth of rainfall on the catchment equaling or exceeding x. In this step we could account for known variations in storm characteristics with location. For example, the average rainfall could be adjusted to account for the location of the grid center. If storm orientations are known to vary throughout the region, the storm could be rotated to conform to the characteristic orientation of the site. Alternatively, one could rotate the storm to several possible orientations and compute a weighted average depth based on the observed frequency distribution of orientations at the site. Next, let $m_i(x)$ be the number of storms which produce an average rainfall x or more on the catchment, when centered at the ith grid center. Then, for any point (y,z) in the ith square, $p[x,(y,z)]$ can be estimated by

$$\hat{p}[x,(y,z)] = m_i(x)/m.$$

Finally, $\hat{p}(x)$ can be estimated by

(4.2) $$\hat{p}(x) = (m/N)(1/A_h)\Sigma[m_i(x)/m]\Delta A$$

or, simplifying,

(4.3) $$\hat{p}(x) = (\Delta A/NA_h)\Sigma m_i(x).$$

Alexander's Method

Alexander (1963, 1969) proposed a storm transposition method which is similar in concept to the one presented above, but which makes a key simplification to facilitate the computation of $\hat{p}(x)$. In particular, Alexander defines a storm as having a fixed area $\bar{A}_s$, determined as follows.

Let $A_s(y,z)$ be the storm area required to cover the catchment if the storm were centered at the point (y,z). In calculating $\hat{p}(x)$, Alexander fixes the storm area to equal $\bar{A}_s$, the average value of $A_s(y,z)$ for all points *in the catchment*. Thus

$$\bar{A}_s = \frac{1}{A_c} \int\!\!\int_{A_c} A_s(y,z) dy dz.$$

Note that by averaging over points in the catchment Alexander is ignoring the possibility that a storm centered outside the catchment might produce the desired rainfall x. Furthermore, rather than actually performing the integral, Alexander recommends using a nominal value of

$$\bar{A}_s = 3A_c.$$

Based on this definition of a storm, Alexander proceeds as follows. For a particular depth of rainfall, let $m(x)$ be the number of storms with mean rainfall equalling or exceeding x (over area $\bar{A}_s$). Then $p(x)$ is estimated by

$$\hat{p}(x) = (A_c/A_h)[m(x)/N].$$

This result can be derived from (4.3), provided the storm area is fixed at $\bar{A}_s$, and rainfall is assumed uniform over $\bar{A}_s$. With these assumptions

$$m_i(x) = m(x)$$

for all grid centers in A_c. Summing over all grid centers in the catchment,

$$\Delta A \Sigma m_i(x) = A_c m(x).$$

Hence from (4.3),
$$\hat{p}(x) = (A_c/A_h)[m(x)/N].$$

This is the same result obtained by Alexander. Note, however, that $\hat{p}(x)$ approaches zero as A_c approaches zero. This erroneous result follows from the requirement that the storm must be centered in the catchment. For a catchment, which is small compared to most storms, this can introduce a significant error.

Yankee Atomic Electric Company Method

The estimate of $\hat{p}(x)$ developed by Yankee Atomic Electric Company (1984) can be obtained from (4.3) in the following way. Define b_{ij} as an indicator variable such that

1 if the average rainfall over the catchment from the jth storm equals or exceeds x when the storm is centered at the ith grid center;

0 otherwise.

Then
$$m_i(x) = \sum_j b_{ij}.$$

Substituting for $m_i(x)$ in (4.3),
$$\hat{p}(x) = (\Delta A/N A_h)\sum_i \sum_j b_{ij}.$$

Changing the order of summation we get
$$\hat{p}(x) = \frac{1}{N A_h}\sum_j \Delta A \sum_i b_{ij}.$$

In the Yankee Atomic Electric Company Method, the catchment is assumed to be a point. Hence $\Delta A\, \Sigma_i b_{ij}$ is area exceeding x for the jth storm. Denote this area $A_j(x)$. Then
$$\hat{p}(x) = \frac{1}{N A_h}\sum_j A_j(x).$$

Finally, letting
$$\bar{A}(x) = (1/m)\sum_j A_j(x),$$

$$\hat{p}(x) = (m/N)[\bar{A}(x)/A_h].$$

This is the same result obtained by Yankee Atomic Electric Company.

Yankee Atomic Electric Company (1984) provides a documented example of the use of storm transposition to estimate rainfall and flood probabilities. The catchment considered has a drainage area of 400 km². Depth–area–duration data were used to estimate $\bar{A}(x)$. To account for the fact that the catchment had finite area, only data from storms which produced an average depth of at least x over an area of 340 km² were utilized. For the smallest depth considered, 15.2 cm, a total of 22 storms were used. A simple model of the spatial distribution of rainfall was used to estimate exceedance areas for various rainfall depths based on the depth–area–duration data. The value of N was assumed to equal 100 years. The area of the meteorologically homogeneous region was 145,000 km². Hence for a range of x from 15.2 cm up to the highest observed depth which occurred over at least 340 km², a nonzero estimate of exceedance probability $p(x)$ was obtained. The maximum observed 24-hour rainfall depth in this case was 61.0 cm; the corresponding estimated exceedance probability was 10^{-7}.

Other Meteorologic Inputs

In addition to rainfall, runoff models may require other meteorologic inputs. Continuous models must account for evapotranspiration. However, evaporation and transpiration influence flood flows only prior to a rainfall event by changing initial moisture conditions. Hence cumulative values of evapotranspiration over several months are important, and these values are generally well behaved and consistent. In the absence of more complete data, monthly mean values of evapotranspiration can thus be used in the modeling.

The situation is much more complicated in the case of catchments that develop extreme flood flows due to snowmelt, rainfall and snowmelt, or rainfall on frozen ground. These are complicated processes that, while fairly well understood, are difficult to model on a large scale. The key meteorologic variables are air temperature, wind velocity, solar radiation, and atmospheric moisture. Snowmelt runoff rates are further influenced by the initial conditions in the snowpack, particularly by the snowpack water equivalent. Meteorologic conditions causing high rates of snowmelt are estimated for probable maximum flood determinations. Systematic statistical studies have not been conducted for these meteorologic conditions. These studies will be needed if a runoff modeling approach is to be used for catchments for which rain floods are affected by snowmelt and/or frozen ground.

ESTIMATION OF PROBABILITIES

The way in which probabilities are estimated depends on whether a continuous or event runoff model is used and on the probability structure associated with the precipitation input.

Estimating probabilities is relatively straightforward when a continuous runoff model is used, either with historic rainfall data or with rainfall amounts generated from a stochastic model. The runoff model is run with the rainfall inputs and the resulting annual flood peaks are treated as a random sample. If historical rainfall data are used to drive the model, a parametric frequency analysis of the resultant annual peaks can be used to estimate exceedance probabilities. Note that for historical records of about 100 years, a considerable extrapolation is required to define floods of several thousand years return period. In the case of a stochastic model, a virtually infinite series of annual floods can be simulated. Hence, model probabilities can be obtained directly from the simulated empirical cumulative distribution function. Regardless of the size of the simulated sample, however, the accuracy of the resultant model flood probabilities is limited by the statistical sampling errors in the observed hydrometeorological data sets used as the basis for specifying the stochastic model.

When one-dimensional temporal models of rainfall are coupled with very simple runoff models, it may be possible to obtain explicitly the probability distribution of floods. Eagleson (1972) does this with a simple exponential model of rainfall and a kinematic wave model of the runoff process. More recently the geomorphic instantaneous unit hydrograph developed by Rodriguez-Iturbe and Valdes (1979) has been used in this fashion. While such an approach is extremely useful for improving our understanding of the rainfall-runoff process, the great simplification it entails may limit its applicability to the problem of estimating exceedance probabilities of extreme floods.

When an event runoff model is used, estimation of probabilities is complicated by the need to specify antecedent conditions. Common practice is to set antecedent conditions at characteristic values, such as historic means or maxima. Undoubtedly this can lead to large errors in the resulting estimates of flood probabilities. A better approach is to treat antecedent conditions as random variables rather than fixed values. This can be done in several ways, two of which are discussed below.

One approach, which would be applicable for use with a stochastic rainfall model, is to develop a stochastic model of antecedent conditions. This model can then be used to generate antecedent conditions for each storm that is generated by the stochastic rainfall model. In essence, the stochastic model of antecedent conditions makes up for the fact that the runoff model is an event model rather than a continuous one. In a sense, it is just an extension of the rainfall model. [For example, Beven (1975) develops a stochastic model of initial discharge and root-zone storage.] Hence, no changes are needed in the procedure used to estimate exceedance probabilities.

When synthetic storms are used as input to the runoff model, another approach is required. In this case explicit probabilities are assigned to the individual storms. Likewise, probability distributions can be assigned to antecedent conditions. Marino and Bradley (1986) present some methods for doing this. Determination of the exceedance probability of a specified flood discharge requires integration of the joint distribution of storm depths and antecedent conditions which can combine to produce floods equaling or exceeding the specified flood discharge. This integration can be facilitated by using methods developed for analogous problems in structural engineering. [See, for example, Madsen, Krenk, and Lind (1986).] A joint probability approach has also been developed for estimating storm surge probabilities [e.g., Myers (1970); National Research Council (1983a).]

ANALYSIS OF UNCERTAINTY

Estimates of the probability distribution of annual floods obtained by runoff modeling will be subject to uncertainties. Quantification of this uncertainty serves three roles. First, it may enable the analyst to choose among competing methods. Second, it enables the analyst to focus on those aspects of the problem which most affect the final results. Third, an estimate of uncertainty in the final results is useful, if not essential, to the decision-maker using the results.

In the previous chapter we discussed how the uncertainty in empirical flood frequency estimates could be quantified. In the case of a runoff modeling approach, the problem is more complex. Such an approach involves the coupling of various models, both statistical and deterministic. Data, which are subject to both error and random variation, are used to estimate parameters of these models. Furthermore, the models themselves are likely to be sources of significant uncertainty. These three kinds of error—data, parameter, and model—require different methods of analysis. Of these, parameter error is the easiest to quantify, particularly in the context of a statistical model. In most specific situations the analyst can use standard statistical techniques to quantify parameter error. Model error is the most difficult to quantify, and may be the dominant source of error. Estimation of model error requires comparison of model prediction with results from the actual process being modeled, or from results generated by other models. In our problem, real-world results are difficult to obtain; hence we may have to resort to the less desirable approach of model comparison. An analysis of model error can and should be performed in the context of specific applications. It is also desirable to conduct systematic evaluations of models that can be generalized for use in specific applications.

In subsequent sections we consider data, parameter, and model error in the use of runoff modeling to estimate exceedance probabilities of rare floods. This is not intended to be a comprehensive discussion of ways to quantify such errors. Instead we focus on the most likely sources of error, those associated with the rainfall input, the runoff modeling, and estimation of flood probabilities.

Errors in the Precipitation Input

Each type of precipitation input has its own sources of error. With the direct use of station data, the key question is how accurately the available data portray the spatial and temporal distribution of storm events. The answer of course depends on the length of record, the density of the gage network, the availability of continuous data, and the spatial and temporal variability of the storm events. Standard statistical methods can be used to evaluate the uncertainty due to finite record length. Wilson et al. (1979), Beven and Hornberger (1982), and Schilling and Fuchs (1986) explore the importance to runoff modeling of accurately portraying the spatial distribution of rainfall. Troutman (1983) uses a linear regression framework to develop estimates of errors in runoff modeling that can result from spatial variability in precipitation.

Stochastic rainfall models are subject to parameter and model error. Parameter error can be quantified by standard statistical procedures. The quantification of model error requires generic studies of individual types of models, such as the study of temporal models performed by Valdes, Rodriguez-Iturbe, and Gupta (1985).

The use of synthetic storms introduces several sources of error. First, the approach itself is a source of error. Storm data are averaged over time (and possibly space) so that univariate statistical analysis can be performed on storm depths over some critical duration. Then "characteristic" spatial and temporal patterns are given to storm depths of various probabilities. The critical assumption is that reasonable estimates of the probability of a specified flood discharge can be obtained by integrating the joint probability distribution of rainfall depths and antecedent conditions over the domain that results in peak discharges equaling or exceeding the specific discharge. This assumption needs to be evaluated, particularly as regards the fixing of storm duration. Continuous stochastic models of rainfall would provide a means for making such an evaluation. The results could be used to quantify the errors that are possible in specific applications of the synthetic storm approach.

A second source of error in the use of synthetic storms is in the estimation of the probability distribution of storm depths. If storm transposition is used, errors can result from incompleteness of the storm catalog, from the reconstruction of historical storms and from the estimation procedure itself. A generic study of storm transposition methods based on a stochastic rainfall model would be very useful in assessing the overall methodology. If storm probabilities are estimated by statistical analysis of station data, standard techniques can be used to estimate the uncertainty in the probabilities of point rainfall. The error in the probabilities of storm rainfall could be substantially larger. Its estimation would require a generic study, which could also involve using a stochastic rainfall model.

Errors in Runoff Modeling

There has been considerable research on the question of errors in runoff-modeling. Little of this research focuses on flood peaks, and there has been virtually no research on the modeling of floods that are much larger than those for which the runoff model is calibrated. This, of course, is precisely the situation with which we are dealing. There is reason to expect that a well-designed runoff model may perform well for floods which are much larger than those for which it is calibrated. Infiltration modeling during extreme floods tends to become less critical as the magnitude of the storm increases, since losses become a smaller proportion of the storm rainfall. More significant errors are likely to result from the flow routing, particularly if there are complications such as complex floodplain flows, extreme channel erosion, extremely high sediment concentrations, or ice-affected flow. In any case, physically based routing methods should be more reliable than empirical methods.

A comprehensive study is needed to evaluate the errors that are possible when runoff models are used to simulate floods that are much larger than the floods used to calibrate the model. This could be done by modeling historically observed extreme floods, such as those cataloged in Bullard (1986). One problem with executing such a study is that there may be relatively few large floods for which there exists reliable data on both rainfall and streamflow.

Errors in the Estimation of Flood Probabilities

The issue here is to determine how the uncertainties that arise from various sources affect the final calculation of flood probabilities. This is a problem in error propagation, which can be addressed by various methods, such as first order analysis and Monte Carlo simulation. Until strategies for using runoff modeling to estimate flood probabilities have been better developed and sources of error explored, it is not possible to be any more specific about this issue.

CONCLUSIONS

Runoff-modeling techniques are in common use for estimating flood probabilities. There are very few well-documented cases, however, where the use involved floods with exceedance probabilities less than 10^{-2}. Furthermore, in none of these cases was much attention paid to the estimation of uncertainties. Nevertheless, runoff modeling appears to be a potentially useful tool for estimating exceedance probabilities of very rare floods.

In view of the dearth of case studies and the lack of attention to analyses of uncertainty, we cannot recommend a specific procedure which would be widely applicable in all situations. However, a few recommendations are possible, regarding both immediate use and needed research.

Runoff Models

Several runoff models exist which appear to be suitable for estimating extreme flood probabilities. Models that are more physically based should perform better than empirical models for floods much larger than those used in calibration. Future research is needed to evaluate the performance of runoff models in simulating well documented historically observed extreme floods, such as those cataloged in Bullard (1986). It is particularly important to determine whether there exist physical processes which are critical to very large floods but which do not operate at lower discharges.

Meteorologic Inputs

Of the various methods for developing rainfall inputs, we favor synthetic storm methods based on storm transposition. These methods have the potential for making the widest use of data over both space and time. Use of this approach requires that there be a complete record of the most extreme storms for some defined period for the meteorological region of interest. The existing storm catalog varies in completeness and length of record, with the most complete record being of storms which have occurred east of the 105th meridian. This storm catalog needs to be expanded, particularly west of the 105th meridian. This is likely to be an expensive effort. Additional research is needed to define those regions of the country that can be considered meteorologically homogeneous and the time period and storm magnitude for which the storm catalog can be considered complete.

We also encourage research on the regionalization of depth–duration–frequency relationships. Very little has been done in this area, and the potential may be even greater than in the regionalization of flood frequency estimation. Regionalized depth–duration–frequency relationships could provide a useful check on probabilities based on the historic storm catalog.

Stochastic modeling of rainfall should also be developed further. Two applications of multisite rainfall models have been applied to the problem of extreme floods, in one case to a small basin and in the other to a moderately large basin. Significant advances can be expected in stochastic modeling in the near future, given the high level of current research activity. As of this date, stochastic models have not been regionalized, and hence their use in estimating the probability of extreme floods can entail significant extrapolation beyond the range of the data on which they have been calibrated. There is no reason, however, why stochastic models could not be regionalized.

Lastly, there has been very little work done on basins for which extraordinary floods are affected by snowmelt. Clearly much additional work is needed.

Estimation of Probabilities

When a continuous runoff model is used, the estimation of exceedance probabilities is relatively straightforward. The difficult case is the use of an event model, which necessitates the specification of antecedent conditions. Additional research is needed to develop and evaluate methods for doing this.

Estimation of Uncertainties

The major sources of errors in using runoff modeling to estimate exceedance probabilities of extreme floods lie in the selection of the rainfall input and in the runoff modeling. Research is needed to quantify these errors.

5

Data Characteristics and Availability

INTRODUCTION

Estimates of magnitudes and probabilities of extreme floods, to be meaningful, must reflect and be consistent with hydrometeorological conditions at the sites of interest. These conditions are represented by various types of hydrometeorological data. Different estimation methods may require different types of data or may use the same types of data in different ways. The purpose of this chapter is to identify the major types of data that may be relevant to the estimation of extreme flood probabilities and to describe any special characteristics of the data that may affect their usability for this purpose.

The principal types of data discussed are streamflow and rainfall data. Streamflow data are used directly by a variety of statistical procedures for estimating extreme flood probabilities, as discussed in chapter 3. Streamflow data also are needed for calibration and verification of the rainfall-runoff models described in chapter 4. Rainfall data can be used in various ways in conjunction with rainfall-runoff models as a surrogate for or supplement to streamflow data.

Although strictly speaking a category of streamflow data, paleohydrologic flood data is discussed in a separate section. This is because paleoflood determinations involve a number of observational and analytical techniques not generally used in the gaging of contemporary streamflows. Also, paleoflood determinations provide more information than just the magnitudes of ancient floods. Under favorable conditions, the paleohydrologic record may furnish information to the effect that the record shows all occurrences of floods greater than some threshold. This or similar auxiliary information is needed for meaningful statistical interpretation of the paleoflood record.

Several other types of data also are important in assessing flood magnitudes and probabilities. Basin-physiographic factors such as drainage areas and soil types are needed for rainfall-runoff modeling and may be helpful in interpreting streamflow records. Other types of meteorological data, including data on evaporation, snow accumulation, and temperature, are necessary for describing important features of the rainfall-runoff relationship, and are discussed briefly.

The principal focus of this report, however, is on the frequency of occurrence of extreme flood events, and not on the mechanisms of flood formation or rainfall-runoff relations. Thus the emphasis of this chapter will be on streamflow and rainfall, the principal variables whose values reflect the occurrence or nonoccurrence of extreme floods, and particularly on the ability of available streamflow and rainfall data sets to meaningfully represent the frequency of occurrence of extreme flood events.

Streamflow and precipitation are complex multidimensional processes. They exhibit strong and complicated patterns of variability in both space and time. It has not been possible to directly measure, record, and store enough data to represent these processes in all their natural detail. Various forms of selection, sampling, and aggregation have been required in order to obtain usable data at an affordable cost. Conflicting requirements for various data uses sometimes require compromises in sampling and observational conditions. The sampling plans and observational procedures thus may affect the usability of a data set for estimation of extreme flood frequency. For this reason relevant aspects of observational procedures and statistical sampling plans for various data types are discussed.

Systematic Versus Nonsystematic Records

For purposes of flood-probability estimation, flood-flow and rainfall data sets can be grouped into two broad categories: systematic and nonsystematic. These categories differ in their ease of statistical interpretation. The essential distinction is that systematic records provide complete records of the occurrences of well-defined classes of events, whereas nonsystematic records define only the occurrences of particular events. The entries in a nonsystematic record are not defined in terms of their relationship to a larger class or population of events; they are merely listed as unique events. Thus, there is no guarantee that a nonsystematic record includes all occurrences of any particular class of events or that absence of an event from the record implies that the event did not occur. Thus the systematic record can be interpreted directly in terms of relative frequencies of occurrence of events whereas the nonsystematic record cannot be so directly interpreted.

Systematic records, as their name suggests, are collected under relatively uniform, consistent, and standardized observation procedures. Although the nature of the conditions of recording may affect the character of the record, the effect of the conditions of recording is known and can be taken into account in the hydrologic interpretation of the record.

For example, the simplest kind of systematic record is an uninterrupted annual flood series or annual-maximum rainfall series, the beginning and ending of which are unrelated to any flood or rainfall event. Each year's maximum flow rate or rainfall depth is recorded, whether or not it is of noteworthy magnitude and even whether or not it might be considered a hydrologically significant event. Another example of a systematic record is a partial-duration series, which is documented in such a way as to ensure that the absence of above-threshold peaks in a year's record implies that no above-threshold peaks occurred during the year. Knowledge about the characteristics of the gage and the protocol of recording are necessary for correct hydrologic interpretation of the partial-duration record. A similar example is a censored annual flood series, in which the magnitudes of only the above-threshold annual peaks are recorded, along with the information that all unrecorded annual peaks during the record period were below the threshold.

In each of these examples, the systematic record was defined in terms of a well-defined class of events. The recorded variables have straightforward and clear-cut representations in terms of random variables associated with repeatable statistical experiments. In each case the description of the systematic record has a natural correspondence with a statistical model as defined in chapter 3.

In contrast to the systematic records, nonsystematic records are not collected under a defined protocol. The collection and preservation of the nonsystematic record depends on undefined auxiliary events, such as the interest or leisure of the observer, in addition to the character of the event of interest. Such records include anecdotal historical accounts from newspapers and journals and personal recollections of long-time residents. Many kinds of paleohydrologic data might have to be included in this category as well, unless the conditions that caused the marks to be made and preserved were known and understood. An example of a nonsystematic record is as follows: "the following noteworthy floods on ____ River were reported in the ____ City newspaper: 1896, 30,000 cfs; 1907, 37,000 cfs; 1924, 45,000 cfs; 1939, 46,000 cfs. The continuous record began in 1943." No information is given as to the basis for observing or recording the peaks from before the continuous record period. There is no basis for associating this record with any particular statistical model.

The essential characteristic of the nonsystematic record thus is that the recipient does not know the system governing its collection. The nonsystematic record may be converted into a systematic one when it is possible to provide additional information that does define the system. Thus if it were known that the peaks listed above were all of those that exceeded 25,000 cfs since 1887, the nonsystematic record would become a complete censored sample that could be analyzed by various historical adjustment procedures. On the other hand, if it were known that the 1896 peak was the largest since 1887 and that the others were successive record-breaking peaks, the record again would become systematic, but would be governed by a different system and would have to be analyzed by more complicated statistical analyses; the simple censored-sample analysis would not be appropriate. An essential prerequisite for statistical interpretation of nonsystematic records thus is the collection of additional information or interpretations that would enable the records to be associated with a statistical model.

Nonsystematic records may provide excellent information about occurrences of events of unusual magnitude, but often do not provide unequivocal evidence about nonoccurrence of such events. Such information may be very useful for defining the potential for future occurrence of extreme events, but does not provide a basis for arguing that larger events cannot occur or for estimating the probability of occurrence.

Statements about flood risk ultimately boil down to assertions about the relative proportions of occurrences and nonoccurrences of events. These assertions ultimately have to be based on experiences of occurrences and nonoccurrences. Systematic records, by their mode of collection, automatically provide this information. Nonsystematic records do not provide the needed information about nonoccurrences, and thus are not so readily usable in flood-risk assessment.

STREAMFLOW DATA

Availability of Streamflow Data

Streamflow data of various kinds are collected and stored by a variety of federal, state, and local agencies. To promote public access to their data, many of these agencies cooperate with the U.S. Geological Survey (USGS) in operating a National Water Data Exchange (NAWDEX). Water-data producers supply information to NAWDEX about locations at which data are collected, the types of data collected, the periods of time for which data are available, and procedures for obtaining copies of the data. NAWDEX, through its assistance centers throughout the country, uses this information to help water-data users identify and acquire relevant available data.

For most data users, the most convenient and efficient points of contact for obtaining streamflow data are the NAWDEX assistance centers. These centers are located in most USGS state and district offices. These offices are located near the capitals or state universities of most states. The locations of NAWDEX centers also can be obtained from the USGS Public Information Office in Washington, D.C. In addition, USGS district offices not only can provide assistance in using NAWDEX to locate and retrieve water data but also can help the user understand the local hydrologic and hydraulic conditions and assess the adequacy and reliability of the available data.

USGS itself collects, processes, publishes, and stores streamflow records for over 11,000 currently active sites nationwide. Most of these records are published in an annual series of USGS Water-Data Reports for each state. These reports are available in many major technical libraries and from the National Technical Information Service. These reports provide a useful overview of hydrologic conditions, data-collection activities, and availability of data in the state. In addition, most of these records are stored in the computerized files of the National Water Data Storage and Retrieval System (WATSTORE). Records for perhaps another 30,000 inactive streamflow sites (an exact count is not available) also are kept on file in the WATSTORE system. Some data collected or processed by other agencies also are stored in WATSTORE; conversely, some other agencies maintain their own water-data management systems that contain data obtained from WATSTORE.

Types of Streamflow Data

USGS maintains three principal computerized files of streamflow data: the peak-flow file, the unit-values file, and the daily-values file. These files contain time-series data with different time steps and different statistical-sampling relationships to the continuous streamflow process. Two other files, the station-header file and the streamflow-basin characteristics file, contain information about the geographic location at which the data were collected. The files were designed to support routine computation, storage, and retrieval of gaging-station data and to support certain types of routine analyses and displays of gaging-station data. The organization and format of the files reflect these intended uses. The primary ordering of all the files is by gaging-station identification number and secondarily, within the station, by water year. Lists of gaging stations within specified geographical areas may be obtained by appropriate searches of the station-header file. Instructions for searching and retrieving these files are given in the WATSTORE User's Guide (Hutchison, 1975; Lepkin et al., 1979), although the most efficient way to identify and obtain the relevant data is with the help of the NAWDEX assistance office in the state where the study area is located.

Time-Series Data Files The peak-flow file contains records of annual maximum instantaneous flood-peak discharges and stages, and their times of occurrence. This file is used primarily to support flood-flow frequency analyses of annual-flood data at gaging stations. These at-site results are correlated with drainage basin characteristics to develop regional flood-frequency relations that are published in separate reports for each state. In addition to the annual maximum values, the file also contains, for many stations, values of instantaneous peak discharges, stages, and times of secondary peaks within the water year that were less than the annual maximum but exceeded a defined flood-base threshold (the so called partial-duration series). Data qualification codes are provided in the file to identify historic (nonsystematic) records and records affected by other nonstandard measurement conditions or hydrologic conditions. The file contains records for about 21,000 locations nationwide, with an average of about 22 years of record per site. Average record lengths for most states are in the range of 18 to 27 years. The areal density of gages is about 7 per 1,000 square miles as an average for the entire conterminous United States; about half the states have gage densities between 5 and 10 per 1,000 square miles.

The available flood-peak data do not everywhere have the temporal and spatial coverage required for developing regional flood-frequency relations. In one attempt to alleviate this deficiency, rainfall-runoff models have been used to transform long-term rainfall records into synthetic flood-peak records (Dawdy et al., 1972; Hauth, 1974).

The WATSTORE unit-values file was designed to hold rainfall and discharge data for selected storm events for use in calibrating rainfall-runoff models and for generating synthetic long-term flood series. The file also contains some water-quality data for use in urban-runoff and land-use studies.

In contrast to the peak-flow file, the unit-values file contains complete storm period discharge hydrographs and rainfall hydrographs, but only for those few sites for which rainfall-runoff models have been calibrated and for only as many storms as were needed to calibrate the model. Only about 4,000 sites are represented in the file, and over half of these are located in just 10 states. The data are recorded as time series with uniform time steps ranging from 5 minutes to 1 hour. The time step is chosen for adequate resolution of hydrographs and tends to be directly related to basin size. Most of the basins represented in the file are quite small; for most of them, rainfall and runoff both were measured at the same location at the mouth of the basin. For convenience of use in generating synthetic flood series, the file also contains selected storm-period data obtained from the National Weather Service for selected long-term rainfall stations.

The third and largest major national water-data file in the USGS WATSTORE system is the daily values file. The file provides essentially continuous records of streamflow, rainfall, evaporation, temperature, stage, and other parameters on a calendar-day basis. In addition to streamflow data, various types of ground-water data and water-quality data also are present in the daily values file. In contrast to the unit-values records, which generally cover only storm periods, the daily values records provide essentially complete coverage of the entire year (or scheduled seasonal periods) for periods of several years or many years. Because the data are given at calendar-day time steps, they may not have adequate resolution for modeling flood peaks on smaller streams. These data are useful, however, for modeling daily-flow flood hydrographs and 1-day and multiday flood volumes. They also are useful for modeling the interstorm drainage and drying processes that determine the antecedent conditions at the beginning of storm periods.

Supplementary Data Files In addition to the three major time-series files are two files that contain supplementary geographic information about the data-collection sites. The station-header file, mentioned previously, contains a record for each site for which data are stored in the WATSTORE system; this amounts to more than 380,000 sites, of which approximately 70,000 are surface-water sites. Information relating to flood flows has been collected at only about 40,000 of these sites. Information in the station-header record includes latitude and longitude, state and county codes, hydrologic unit code, drainage area (contributing and noncontributing), and elevation. More detailed information for approximately 16,000 sites is contained in the streamflow-basin characteristic file. This file was designed to support various studies of correlation between streamflow characteristics and drainage-basin characteristics. Among the characteristics represented in the file are flood-peak discharges and volumes for various return periods and durations; basin latitude, longitude, and elevation; basin length, width, and orientation; drainage area; basin and channel slopes; percentages of forest and wetland; and mean annual precipitation and 2-year, 24-hour rainfall intensity. Not all characteristics are available for all basins. The streamflow characteristics, however, can be readily determined from the basic data in the peak flow and daily-values files.

Not all available data are stored in centralized computer files. Many flood-frequency and rainfall-runoff studies require unique statistical and hydrologic computations that are not provided by the computer software packages associated with the centralized water-data file system. It frequently happens that, for purposes of computation, the data are compiled and stored in the format required by the computer software being used. When the projects are complete, the data are published in either data reports or interpretive reports and are sent to the archives without ever entering the centralized computer files. This is more likely to happen to basin-characteristic data than to field-collected rainfall and runoff data. An extensive bibliography of USGS flood-frequency reports is given by Thomas (1986a); many of these reports contain detailed tabulations of flood and basin characteristics. However, the most efficient way to identify and obtain such data is with the assistance of local USGS NAWDEX user-assistance personnel, who will be familiar with the identity, location, and status of such data sets.

Finally, all USGS districts maintain extensive files of flood-discharge determinations. These files include both direct current-meter measurements and indirect measurements by methods such as the slope-area method. Most of the measurements are made at active gaging-station sites, but significant numbers are made at sites of discontinued gages and at miscellaneous ungaged sites. The measurements at active gaging stations are used primarily to develop, extend, and verify the stage-discharge ratings for the stations. These measurements have no direct applicability to estimation of extreme flood probabilities. Most of them refer to low and medium flows. Information on flood flows at gaging stations is contained in the peak-flow file.

Flood measurements at miscellaneous ungaged sites are made to document the occurrence, extent, and magnitudes of extreme floods. These measurements are made to supplement the areal coverage provided by the regular gaging stations. These records are nonsystematic in the sense discussed above. They cannot readily be related to any statistical model. Measurements are made only in response to known extreme flooding in the area and are intended to provide evidence of the ultimate potential for extreme flooding. Flood damages and local interest in obtaining information also are considered when deciding whether to make these measurements. These records thus may be used to develop flood envelope curves (Jarvis, 1926; Creager and Justin, 1950; Crippen and Bue, 1977; Costa, 1985), but do not provide any direct evidence as to the frequency of occurrence of extreme floods. The measurements themselves usually are published either in annual data reports or in special flood-documentation reports. Thomas (1986b) has compiled a bibliography of USGS flood-documentation reports. Because the arrangement and management of the miscellaneous-measurement files vary from district to district, the assistance of local USGS personnel should be obtained to identify and retrieve relevant information.

General Principles of Streamflow Data Collection and Computation

Streamflow data are collected by three basic methods: direct measurements, indirect measurements, and calibration-curve methods. Direct measurements of streamflow are made by direct measurement of the velocity of the water and the width and depth of the channel; the discharge is computed from its definition as the product of the directly measured factors.

Because direct measurements often cannot be obtained during the occurrence of peak flows, extreme flood peak flow sometimes must be determined by indirect methods. Indirect discharge determinations use various hydraulic formulas, most commonly the Bernoulli and Manning formulas, to calculate flood-peak discharges based on postflood surveys of high-water marks and channel dimensions and on field estimates of roughness coefficients and other relevant hydraulic factors (Benson and Dalrymple, 1967; Barnes and Davidian, 1978).

Direct and indirect measurements both provide flow data only at discrete instants of time. They are not practical for collection of continuous records throughout an interval of time. For this purpose, various calibration-curve methods are used. Most commonly, a continuous record of water stage or gage height is collected by means of a counterweighted float or a gas-pressure manometer connected to a suitable recording apparatus. The recorded gage heights are applied to a stage-discharge rating curve or calibration curve to obtain the corresponding discharges. The rating curve is established by correlating direct and indirect discharge measurements with their corresponding gage heights. Such a correlation is possible at suitably chosen sites where both cross-sectional area and velocity are definable functions of stage (or of flow depth). At some sites more complicated calibrations are needed, involving factors such as rate of change of stage, variable water surface slope, or velocity indices, in addition to stage, but these factors frequently diminish in importance under flood-peak flow conditions and so need not be discussed here. The water stages commonly are recorded on computer-readable media, the stage-discharge conversions commonly are done by computer, and the results are stored directly in computer files.

Methods of streamflow data collection and computation are described briefly by Linsley et al. (1982) and in detail by Rantz (1982), Brakensiek et al. (1979), and Office of Water Data Coordination (1977).

Accuracy of Streamflow Data

Accuracy of flood-flow data depends in part on the method by which the flow was determined. In general, the number and severity of the error sources increase with the degree to which the data depend on interpretation and calculation as opposed to direct physical measurement. Principal sources of error include loss of data during periods of extreme flood flow, inaccuracies of direct and indirect flow measurements, inadequate definition of stage-discharge relations for extreme floods, and inadequate interpretation of available data. Errors in flood-flow data may be large enough to distort the statistical properties of the flood series (Potter and Walker, 1982).

Direct current-meter measurements are affected primarily by problems in obtaining representative mean velocities, problems in measuring depths and velocities using sounding lines in swift and debris-laden currents, and problems in completing the measurement quickly enough for it to be meaningful during rapidly varying flow conditions. The USGS rates its current-meter measurements as excellent if their errors are believed to be less than 2 percent, good if 5 percent, and fair if 8 percent (Rantz et al., 1982, p. 178 ff).

Indirect measurements are affected by problems in finding consistent high-water marks to define the water surface profile, problems in analyzing postflood evidence to determine channel and flow conditions including scour during the peak flow, problems in determining the hydraulic laws that governed the flow, problems in estimating judgmental coefficients such as Manning's n, and problems in defining flat water-surface slopes with sufficient accuracy (Benson and Dalrymple, 1967; Jerrett, 1985). The USGS rates its indirect measurements as good if errors are judged to be less than 10 percent, fair if less than 15 percent (Benson and Dalrymple, 1967).

Accuracy of calibration-curve methods is affected by errors in stage measurement, and by errors in the stage-discharge rating curve. Stage-measurement errors can be caused by surge, drawdown, stagnation, or plugging or fouling of intakes or orifices. Many of these effects are most severe at the high velocities associated with high flood flows. Flood damage sometimes may result in complete loss of the stage record during the flood peak. These errors may be mitigated to some extent by conscientious use of high-water marks to verify or supplement recorded peak stages. Errors in the stage-discharge rating result from the combined effects of all errors affecting the measurements of stage and discharge upon which the rating is based.

Although the measurement errors tend to be averaged out during the construction of the rating curve, this effect is largely offset by the scarcity of measurements at extreme flood flows. Most ratings need to be extrapolated to accommodate extreme flood peaks. The best-available rating-extrapolation methods use hydraulic principles, coefficients, and data that are comparable to indirect discharge measurements. Ratings extended by step-backwater calculations may have errors of 15 or 20 percent (Cook, 1985). The USGS rates its published streamflow records as excellent if about 95 percent of the daily discharges are judged to be within 5 percent of their true values; good, if within 10 percent; fair, if within 15 percent; and poor otherwise (Rantz, 1982, p. 613).

Accurate determination of flood magnitudes is only one aspect of the general question of the accuracy of flood data. Other issues that relate to the usability of flood data for estimation of flood probabilities include spatial and temporal correlations and the statistical characteristics of the sampling plan governing the data-collection program. These issues apply to rainfall data and paleohydrologic data as well as to streamflow data and thus are discussed separately at the end of the chapter.

RAINFALL DATA

Precipitation data useful for determining extreme flood events include the following:

- Point observations (systematic)
— Continuous daily and hourly observations of rain and snow at gages.
- Areal observations (Some of these data, in both of the categories below, can be classified as systematic.)
— Weather radar observations of rainfall intensities.
— Satellite observations of distribution or persistence of clouds. (Quantitative estimates of rainfall provided by satellite data have not yet been used in studies dealing with extreme flood events.)
- Extreme-event observations (generally nonsystematic)
— Some possible sources: "bucket surveys," diaries, histories, disaster surveys.

For each of these data types, different methods for observing, processing, and archiving the data are used. The errors associated with each type of observation vary widely. In the paragraphs below we describe observational networks, data acquisition, and available information about the errors associated with the data.

Availability of Rainfall Data

Rainfall and other meteorological data are collected for various purposes by many federal, state, and local agencies and by private organizations and individuals. To make this data more readily available, the National Oceanic and Atmospheric Administration (NOAA) operates the National Environmental Data Referral Service (NEDRES) within the National Environmental Satellite Data and Information Service (NESDIS), Washington, D.C. NEDRES is a publicly available service that identifies the existence, location, characteristics, and availability of all types of environmental data sets. NEDRES itself does not provide the actual data but does enable the data requestor to make contact with holders of appropriate data sets. For hydrologic data relating to extreme floods, the NEDRES activities are coordinated with the National Water Data Exchange (NAWDEX) of the U.S. Geological Survey.

The most comprehensive source of precipitation data (and other climatic data) for use in extreme flood studies is the National Climatic Data Center (NCDC) in Asheville, North Carolina. The NCDC archives and publishes various kinds of weather and climatic data collected by the National Weather Service (NWS) and by other NOAA-supervised stations, including the Cooperative Observer Network. Available data are organized into a large number of predefined data sets described in the NCDC catalog, "Selective Guide to Climatic Data Sources" (Hatch, 1983). Additional specialized searches and tabulations can be obtained if necessary by special arrangement with NCDC. The published data sets that seem most likely to be of use in the estimation of extreme flood probabilities are described briefly in the following paragraphs. These data are published both as computer-readable files and as serial publications that are available for reference in many major technical libraries.

Daily precipitation records, obtained primarily from the NWS Cooperative Observer Network, are published for approximately 8,500 currently active stations nationwide. Archived observations from before 1900 to the present are available. The data are published on a current basis for each state by NCDC in the monthly and annual *Climatological Data* series listed in the NCDC catalog (Hatch, 1983). The data are published in computer-readable form as Summary-of-the-Day and Summary-of-the-Month Co-op Element Files. In addition to precipitation data, these files may contain data on several other weather parameters, including snowfall and snow depth, maximum and minimum temperatures, and evaporation. The complete nationwide data set occupies more than 100 reels of magnetic computer tape. It is possible to select only those states, stations, and parameters of interest. Not all parameters are available at all stations. The locations of stations and the various available types of data normally collected are shown on maps in the *Climatological Data* publications.

Hourly precipitation data are published by NCDC for over 5,550 currently operational stations nationwide. These records are obtained primarily from NWS and Cooperative Observer stations equipped with continuous raingage charts or punched-paper-tape recorders. Hourly data collected during the period from before 1948 to the present are available. A smaller amount of data observed at 15-minute intervals is available for the period from 1971 to the present. These data are published on a current basis for each state except Alaska in the NCDC's monthly and annual *Hourly Precipitation Data* series listed in the NCDC catalog (Hatch, 1983). The data are available in computer-readable form in the Hourly and 15-Minute Precipitation Element Files. The complete data set occupies approximately 20 computer tape reels. Data on magnetic tape can be ordered by state, station number, and date. It is possible to select only those stations of interest. Maps in the *Hourly Precipitation Data* publications show the locations of stations collecting hourly data.

Daily, hourly, and more frequently observed precipitation data also are collected by a wide variety of other governmental, institutional, and private observers not included in the NWS Cooperative Observer Network and not published by NCDC. The U.S. Agricultural Research Service (ARS), for example, operates several highly instrumented research watersheds with dense coverages of raingages and runoff gages (Thurman and Roberts, 1987). The ARS data and many other non-NCDC data sets are indexed in the NEDRES and NAWDEX data-referral services. Many of the remaining observations are used solely for current operational purposes and are not archived for future reference.

A large number of stations, approximately equal to those in the published hourly precipitation network, report through the Geostationary Operational Environmental Satellite (GOES) Data Distribution System. About one-half of these (over 2,000) report precipitation. GOES-reporting stations are owned by several organizations, primarily the U.S. Army Corps of Engineers and the U.S. Geological Survey. The stations are unattended automatic data collection platforms (DCP's) that generally make observations every hour and report every 3 or 4 hours. Reporting times are offset so that the reporting rate is spread evenly throughout the 3- or 4-hour cycle. Some of these stations also report data whenever rainfall intensities exceed a preset value. The data can be accessed in real time by a direct-readout down-link, through arrangement with NOAA or the station owner. Data archiving and quality control are under the control of the owner of the DCP. Many of these same stations report in the NOAA publications of hourly or daily data mentioned previously.

Accuracy and Interpretation of Precipitation Data

Several characteristics of precipitation data may affect its usability and proper interpretation in extreme flood probability studies. These characteristics relate both to the accuracy of the recorded values at the measurement point and to the extent to which the measured values are representative of the past and future occurrences of precipitation in the area.

Care must be used in applying and combining the data. In runoff studies, for example, it is important to note that precipitation data published by NCDC do not distinguish between liquid precipitation and water equivalent of snow. Precautions must be taken when part of the precipitation is snow. The need for caution in relating runoff to precipitation events will be discussed under the section on snow data.

Similarly, the classification "daily precipitation data" generally is construed to mean precipitation that is accumulated in a gage over a 24-hour period and then measured. The accumulation period, however, is not necessarily tied to a calendar day. The NWS cooperative network stations do not all take observations at the same times. Some stations take observations in the evenings, some in the mornings, and some at other times. For the NWS cooperative network, reporting times are published for each state in the *Annual Summary of Climatological Data* series and also are available by computer. A difference in observation times can cause the same storm to appear to occur on different days. For example, a storm occurring at noon on January 1 would be observed as January 1 by an evening observer but as January 2 by a morning observer.

In addition to these fundamental problems of interpretation, numerous human, mechanical, and environmental errors affect precipitation data. Measurement procedures and many of the more common error sources are described briefly by Linsley et al. (1982) and in more detail by references cited below.

Wind is perhaps the main cause of errors in all point precipitation measurements. This error is especially noticeable with snow. Errors caused by wind are described by Peck (1972a,b) and Larson and Peck (1974). Changes in the station environment that occur with time can also cause errors, e.g., urbanization, changes in observers, removal or regrowth of vegetation, etc. Tests for such changes and corrections to restore intergage consistency are discussed by Chow (1964), Hydrologic Research Laboratory Staff (1972), and Sevruk (1982).

To help minimize these and other errors, stations submitting data to be published by the NODC are expected to operate standard equipment and follow a standard procedure, with observations taken at standard times. These standard procedures are described in *Weather Bureau Observing Handbook No. 2* (WB-ESSA, 1970), currently (1987) undergoing revision. Stations are visited by NWS personnel once or twice a year to be serviced and to verify compliance with standard operating procedures. The period of record varies from station to station but is generally indicated in the published records.

In addition to the procedures described in *Observing Handbook No. 2*, there are very general descriptions of methods for observation of precipitation contained in the *National Handbook of Recommended Methods for Water Data Acquisition* (Office of Water Data Coordination, 1977) and the *Field Manual for Research in Agricultural Hydrology* (Brakensiek et al., 1979). A brief description of precipitation measurement methods also is given by Linsley et al. (1982). There are also many non-NWS supervised networks that maintain some standards for equipment and observations. Some networks, however, accept almost any available data, regardless of quality. Thus data quality from available non-NWS networks covers such a broad spectrum that it is difficult to generalize about its accuracy or precision.

Special Precipitation Networks

The distribution of official observation stations is neither uniform nor particularly well suited to defining the details of the distribution of precipitation in storms. In many instances, intense precipitation amounts may go entirely unobserved. Sometimes numerical estimates of amounts of rainfall in miscellaneous containers such as buckets and bowls are collected from residents of areas receiving heavy rain. These unofficial "bucket surveys," made after some unusual events, may disclose amounts that allow isohyetal analysis to point to a likely maximum. Such nonsystematic data may serve to enhance systematic rainfall records.

From time to time special dense networks of rain gages are operated by various agencies, frequently for research in areal distribution of precipitation. Although useful for this purpose, the data collected at these sites often cannot be considered indicative of severe storm probabilities because the data often are collected only on an ad hoc basis during significant storms, without consistent statistical sampling criteria. In addition, some networks collect data routinely for operational purposes but do not archive the data. Summaries of special networks have been made by the American Geophysical Union (AGU, 1965) and by the American Society of Civil Engineers (Tucker 1969, 1970a,b). These summaries are not complete, however, and numerous other networks exist. Many television stations, for example, sponsor local Weatherwatchers networks. A comprehensive survey of such data has not been made and would be difficult to make. The NEDRES and NAWDEX data referral services might be useful for identifying some networks having these types of data.

The Storm Rainfall Catalog

Storm Rainfall in the United States is an unpublished catalog of storms collected jointly by the U.S. Army Corps of Engineers and the National Weather Service. Assembly of this catalog was begun by the Corps in 1937 because of their interest in accumulating a comprehensive set of extreme precipitation data. The data have been used primarily to develop estimates of probable maximum precipitation (PMP). Briefly, all available data for each individual storm were assembled and mass curves showing the time distribution of rainfall at each station were constructed. These mass curves aided in setting the time distribution of the storm. A depth–area–duration

(D–A–D) matrix of average precipitation was computed (Weather Bureau, 1946).

The principal results of these studies are illustrated in Figures 19 and 20. It is cautioned that the D–A–D data (Figure 19) are not a chronological sequence of rainfall, but the largest amounts for the incremental periods indicated. It is also true that these data are nested in time such that the 12-hour amount is contained in the 18-hour amount, etc. Isohyetal maps (Figure 20) usually are available only for total storm precipitation. The records of these studies are stored at the Directorate of Civil Works, Office of the Chief of Engineers in Washington, D.C. We emphasize that the sample of completed storms is not a complete record of significant storm events either in time or in location. Models requiring complete data cannot be applied until these inadequacies are remedied. Attempts to use incomplete data will cause the results to be questionable (see research recommendations, chapter 6).

Currently, a total of 563 storm studies have been completed. The spatial distribution of the storms is shown in Figure 17 in chapter 4. Very few storms in the mountainous western states have been analyzed for inclusion in the Storm Catalog because of problems inherent in orographic regions.

Over 400 of the storms in the catalog occurred between 1900 and 1940. It seems likely that all the pre-1900 storms were outstanding events that required analysis despite limited data. Similarly, most post-1940 storms were selected because they exceeded previous storms or occurred in locations or seasons not previously analyzed. As the catalog accumulated, a "background" of extreme storms was established, and only those additional storms that exceeded the other storms were processed. The "threshold" for including new storms thus varies from region to region.

The storm analysis is currently a manual process that can take more than a year to complete for the larger storms. The number of storms studied has been restricted by limited funds. As a result, many extreme storms, especially those that occurred after the 1940's, have not been analyzed. Thus, the storm record presented in the catalog is not statistically complete.

There is considerable variability in the data sets used to develop the analyses. Almost all the first storm analyses were based on archived data. The oldest storm studies thus were usually based on data from fewer stations than the more recent studies. Analyses of some storms that occurred after 1940 may have benefited from

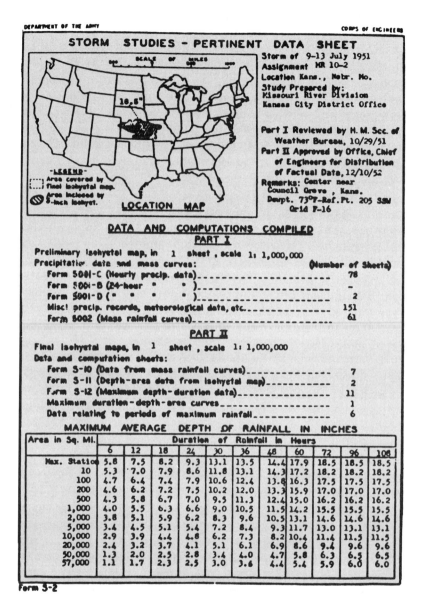

FIGURE 19 Example of completed United States storm-study results—storm location map and maximum depth–area–duration data for the storm of July 9-13, 1951, centered near Council Grove, Kans. From U.S. Army Corps of Engineers, ongoing analyses: 1945 to present.

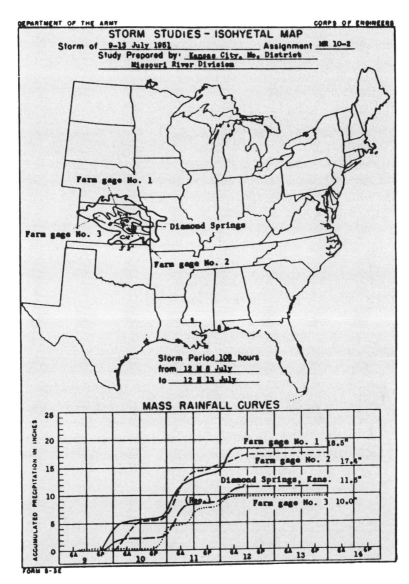

FIGURE 20 Example of completed United States storm-study results—total-storm isohyetal map and mass curves for principal rainfall centers for the storm of July 9-13, 1951, centered near Council Grove, Kans. From U.S. Army Corps of Engineers, ongoing analyses: 1945 to present.

supplemental data, including that from unofficial "bucket surveys." The quality of the resulting analyses is undoubtedly affected by the number and the spatial distribution of the available observation locations. The resulting inconsistencies need to be given proper consideration in any attempted statistical analysis of this data set.

Extremes of depth–area–duration data also have been compiled by Shipe and Riedel (1976). They summarized the greatest known rainfall depths for various areas and durations for 5° × 5° latitude–longitude grid units of the United States. The input data for this summary have been placed on magnetic tape. Additional work in this area was done by Riedel and Schreiner (1980), who compared the greatest observed rainfalls with generalized PMP estimates.

Recently, in 1985 the Bureau of Reclamation began to automate the storm analysis procedure and hopes to reduce the time to complete a study to about 1 month per storm, although some work still has to be done manually. The intent is to provide a means to process a relatively large number of storms in the western United States and thereby improve studies of Probable Maximum Precipitation for that region.

Similarly, the Canadian Atmospheric Environment Service has developed a semiautomated storm analysis system. In this system, computer data files and procedures are used to plot maps showing total storm rainfall depths at gage locations. The maps are interpreted and isohyetal contours drawn manually. The total-depth isohyets then are digitized. The computer apportions the depth by time and area using available recording-gage data and computes and prints depth–area–duration tables and charts. A catalog of over 500 severe storms has been compiled (Pugsley, 1981).

Snow Data

Precipitation falling completely or partly as snow generally cannot be associated with runoff attributed to a single storm. Caution should be taken in using observations of such precipitation to generate runoff from storms. The runoff from solid or mixed precipitation may be significantly reduced by that portion of the precipitation remaining as snow cover in the basin. This situation frequently occurs in mountainous areas. However, especially large runoff can occur from basins with significant snow cover when rains are accompanied by warm temperatures. But on some occasions (fairly deep snow pack and cool temperatures), the snow pack may absorb the liquid precipitation resulting in significantly slower runoff.

Generally, snowfall data are considered a subset of precipitation data. Available snowfall and snow on the ground data are stored on the same magnetic tapes as the daily rainfall data, at NCDC in Asheville, N.C. Another source is the World Data Center A for Glaciology (Snow and Ice), University of Colorado, Boulder. This Center has data different from the hydrometeorological data available at NCDC, e.g., water equivalent of snow obtained from gamma detection flights conducted by NWS on selected flight lines in the upper midwest during the winter.

The primary agencies archiving snow-gage and snow-course data are: Soil Conservation Service (SCS) of the U.S. Department of Agriculture, the Eastern Snow Conference, and the California Department of Water Resources. The SCS data includes observations from a telemetered snow-monitoring network called SNOTEL. Snow data from the eastern United States may be obtained from the Proceedings of the Eastern Snow Conference. Snow data from the west (except for California) are available from the SCS series of publications *Water Supply Outlook and Federal State Private Cooperative Snow Surveys*. Snow data for California are published in *Water Conditions in California* by the California Department of Water Resources. Statistical summaries of extreme snowfall data have been published by Ludlum (1962) and Thom (1957).

Radar Rainfall Data

General

An advantage of this form of data is that radar observations of precipitation can be taken over large regions (radius of up to 85 km) during relatively short periods (minutes). Currently there are two forms of operational digital rainfall data from radar systems. Both are derived from the primary radars employed by NWS.

In the first form, called manually digitized radar (MDR), a radar operator manually digitizes radar echo intensities. The product of such digitization is based on the maximum echo intensity within an MDR box (22 by 22 nautical miles) during the observation period (1 hour). The product does not specify the precise location of the maximum echo within an MDR box.

The second form comes from a system called RADAP II (Radar Data Processor II, previously called D/Radex)(Greene et al., 1983).

It is a fully automated system capable of creating several products, including estimates of rainfall accumulations, for periods as short as 1 hour. The RADAP II system is presently implemented at 11 locations in the country.

In the relatively near future (1990–1992), a new weather radar system called Next Generation Weather Radar (NEXRAD) will replace both of the systems just mentioned. Rainfall estimates in digital format will be provided hourly on a grid approximately 4 km on a side. The quality of these data will be superior to the quality of the data collected by existing systems because the hardware and software designs minimize many error-causing factors.

Other radar systems in the country are operated primarily in a research mode. Two important systems are those operated by NCAR in Boulder, Colo., and the National Severe Storms Laboratory in Norman, Okla. Additional systems are operated by universities, such as Massachusetts Institute of Technology, Texas A&M University, and the University of Chicago. Information on the operation and data archiving policies should be obtained directly from these centers.

MDR Data

MDR data have been archived since November 1973. The reflectivity intensity codes along with the number and type of severe local storms that have occurred during the hour in an MDR box are archived. The archiving is done by the Techniques Development Laboratory (TDL) at NWS Headquarters in Silver Spring, Md., and the data are available through the National Center for Atmospheric Research (NCAR) in Boulder, Colo. The MDR data serve as a basis for the National Radar Summary chart, which shows hourly radar data across the country, annotated with severe storm information. These charts are archived on microfilm at NCDC. These data are systematic.

MDR data are generally continuous and checked for errors. However, the quality of MDR data is a function of the skill of the radar operator. Also, severe weather affects the quality of MDR data. A busy and tired operator is bound to make more mistakes under severe weather conditions. Low resolution, in both space and magnitude, makes MDR data of limited usefulness for flood frequency studies; however, these data could be used to assist in determination of spatial distribution of storms.

RADAP II Data

Routine archiving of RADAP II data began in 1985. The archived data consist of reflectivities for every 10 minutes for as many standard elevation angles (usually four) as necessary to reach the storm top. The data are archived by the Oklahoma Climatological Survey in Norman, Okla., and on an "as time permits" basis by the Techniques Development Laboratory of the National Weather Service. No hourly or longer-period accumulations are routinely archived. Users interested in accumulated rainfall must do the necessary processing themselves. Several algorithms for this task are available from the TDL (W/OSD2) or the Office of Hydrology (W/OH) at NWS Headquarters, Silver Spring, MD 20910.

Radars supplying the RADAP II data can be taken out of the automatic mode whenever the operator desires and in the past have frequently been diverted to other activities during severe storms. For this reason these data cannot be considered systematic. The major problem with quality is occasional anomalous propagation (AP) of the radar signal. Other important contributors are calibration errors, partial beam-filling, and the variability of the relationship between the reflectivity and the rainfall rate. For a more detailed discussion of radar and its uses, see, for example, Battan (1973). In order to partially eliminate errors, it is recommended that radar analysis be merged with rain gage data using procedures such as those described by Crawford (1979) and Krajewski and Hudlow (1983).

Satellite Data

Satellite data considered in this chapter are sensed by instruments on board a satellite and telemetered to the ground for further use. This definition excludes in situ information that is transmitted from one point on the ground to another via satellite.

Those data in most common use are the data from the Very High Resolution Radiometer on board the NOAA polar orbiting satellites and from the Visible-Infrared Spin Scan Radiometers on board the GOES weather satellites. The observations (imagery) from the NOAA and GOES weather satellites primarily consist of reflected visible radiation and emitted thermal infrared radiation. Resolutions range from 1 to 8 km. This imagery can be used to note persistence of clouds over storm areas. The application of these data to rainfall estimation is described by Farnsworth et al. (1984). Access to both archived and real-time data can be arranged through the NOAA National Environmental Satellite and Data Information Service (NESDIS).

Air Temperature

Air temperature is primarily of interest as an aid in differentiating between rain and snow and as the primary observation used to drive snowmelt model simulations. NWS stations in the United States have been recording temperatures in degrees Fahrenheit. Most thermometers used to measure temperature are accurate to within 1 degree. Daily maximum and minimum temperature values are the values most commonly recorded and archived in the climatological data sets available from NCDC, as discussed in the rainfall data section.

Some stations, designated as synoptic stations, observe hourly or 3-hourly surface air temperature values. These observations also are available from NCDC. Upper air temperature soundings from the radiosonde network covering the United States, which may be useful for determining whether precipitation is rain or snow, are available as described in the NCDC data catalog.

Evaporation

Hydrologic models used to estimate runoff from rainfall or snowmelt must account for water lost to evapotranspiration. Transpiration is highly dependent on vegetative cover and therefore quite site specific. Evaporation is measured during months with above-freezing temperatures by a network of pans covering the entire country. These daily data are published in *Climatological Data* and are available on magnetic tape with other daily observations from NCDC. Two NOAA reports on evaporation, one an atlas (Farnsworth et al., 1982) and one a tabular report of United States pan data (Farnsworth and Thompson, 1982) are available.

Other Hydrometeorological Data

There are many networks, both private and public, observing hydrometeorological data. Generally, the data that are easiest to access are those of the NWS networks that report climatic data to NCDC. While NCDC archives data from many sources, it publishes only the data meeting the standards set forth in *Weather Bureau Observing Handbook No. 2*. Most of these data are listed in the publication *Selective Guide to Climatic Data Sources* (Hatch, 1983). This guide includes information on data such as upper air soundings, solar radiation, storms, and storm tracks.

An additional source of hydrometeorologic information is contained in various atlases. These data are important, for example, when using procedures involving transposition of storms from one basin to another. Analysts using such procedures should consider climatic probabilities of storm direction. Climatic norms for many variables, including prevailing winds for selected stations, are depicted in the *Climatic Atlas of the United States* (Environmental Data Service, 1968).

Also of interest are rainfall-frequency atlases, including those of Hershfield (1961), Miller et al. (1973), Miller and Frederick (1966), and Frederick et al. (1977).

PALEOFLOOD HYDROLOGY

General Characteristics of Paleohydrologic and Historical Records

In the broadest sense, paleohydrology is the study of movements of water and sediment before the time of continuous measurement by modern hydrologic procedures. In the study of ancient floods, a distinction can be made on the basis of human observation. Historical flood records involve human observation and documentation of the actual flow events at the time of their occurrence. Because historical records often have been accumulated under a mixture of observational techniques and criteria for determining whether events were significant enough to record, these data require careful evaluation to make them compatible with data derived from modern hydrologic procedures. Paleohydrologic records, on the other hand, are produced by physical processes during the occurrence of the ancient flood. Where this evidence has been preserved, it may be possible to reconstruct the age and magnitude of these paleofloods (Baker, 1985; Kochel and Baker, 1982). Because paleoflood records are produced by deterministic physical processes and tend to be preserved at stable geological sites, they may in some cases be superior to historical records in terms of accuracy and ease of interpretation.

In this report we distinguish between a broad category of paleogeomorphic-based flow data and a specific category of paleostage-based flow data. The reason for the distinction is that in the broader class of geomorphic settings, paleoflow estimates can be achieved only with relatively low accuracy. In certain geomorphic settings, however, paleostage studies can yield accurate determinations of discharge and age for paleofloods.

Paleogeomorphic-Based Flow Data

Many techniques are available to extend flow records into the past using principles of geomorphology and related aspects of Quaternary stratigraphy, sedimentology, and geobotany (Costa, 1978; Costa and Baker, 1981; Foley et al., 1984; Gregory, 1983; Williams, 1984). Among the categories of analysis are regime-based paleoflow estimates, maximum-particle-size-based estimates and paleostage-based estimates. The latter, because of their greater accuracy, will be treated separately.

Regime-based paleoflow estimates (RBPE) involve empirically derived relationships that relate relatively high-probability flow events, such as the mean annual flood or bankfull discharge, to paleochannel dimensions, sediment types, paleochannel gradients, and other field evidence. Such relationships apply to alluvial channels, which adjust their width, depth, sediment transport, and slope to the flood discharge. These variables are related to discharge by regression expressions derived from observations of relationships in modern alluvial channels. RBPE studies have been summarized by Dury (1976), Ethridge and Schumm (1978), and Williams (1984). Because nearly all RBPE relationships apply only to relatively frequent floods, they are of little use in the evaluation of extreme floods.

Maximum-particle-size studies involve regression expressions and theoretical considerations that determine shear stress, velocity, or stream power (SS–V–SP). The paleohydrologic applications assume that the SS–V–SP estimates apply to the maximum floods that transported the particles. Whether based on theoretical (Baker, 1974) or empirical procedures (Costa, 1983), the key data for SS–V–SP studies are paleochannel dimensions, slopes, and maximum particle sizes. A critical assumption is that particles sufficiently large to have been transported near the competence limit of the flood were present in the reach. Church (1978) notes that the error can reach an order of magnitude where local controls on sediment transport are not known. Much better accuracy can be achieved by a combination of procedures (Costa, 1983), but irreducible error remains high and nearly impossible to specify accurately. Although SS–V–SP studies apply to rare, large floods, such studies usually will produce only a single estimate of the largest flood experienced in a given time period.

Paleostage-Based Flow Data

General Principles Paleostage-based flow data are generated from stable-boundary fluvial reaches characterized by slackwater deposition and paleostage indicators (SWD–PSI). Slackwater deposits consist of sand and silt (occasionally gravel) that accumulate relatively rapidly from suspension during major floods, particularly where flow irregularities result in markedly reduced local flow velocities. Under suitable conditions, the tops of these deposits indicate the paleostage or maximum elevation of water surface during the flood. By use of hydraulic calculations involving paleostages at one or more sites in the reach, the channel dimensions, and appropriate hydraulic resistance coefficients, the corresponding peak flow rate can be determined. The hydraulic calculations are essentially similar to those used in indirect discharge determinations based on high-water marks of contemporary floods. What is unique is the methodology used to identify the paleo-high-water marks (paleostages), to determine the ages of these marks, and to establish the paleochronological sequence of flood occurrences at the site.

Stage Discharge Computations After paleoflood stages have been established by methods to be described below, the stage data must be transformed into paleodischarge estimates. This can be accomplished by several hydraulic procedures, including the slope-area method (Barnes and Davidian, 1978; Dalrymple and Benson, 1967) and the step-backwater method (Davidian, 1984; Feldman, 1981). The slope-area method was used in the first SWD–PSI studies (Baker et al., 1979, 1983; Baker, 1983; Kochel and Baker, 1982; Kochel et al., 1982), but more recent work has used the step-backwater analysis (Ely and Baker, 1985; O'Connor et al., 1986; Partridge and Baker, 1987). A significant advantage of step-backwater analysis over the slope-area method is that the step-backwater calculations can be performed independently of the high-water indicator survey to produce a stage-discharge relationship covering a range of stages and discharges. This relationship can be used with a smaller number of more poorly defined paleostages than might be required for a slope-area calculation.

Of prime importance to accurate flow modeling is an accurate characterization of channel geometry. The hydraulic cross sections should be chosen to be representative of the reach between them and to permit proper evaluation of energy losses due to friction, expansion, and contraction. The channel geometry should be defined for a sufficient distance downstream to eliminate profile dependence on the starting downstream stage. Roughness coefficients should be estimated and documented on the basis of field observations. When sufficient and appropriate cross sections are chosen to define the flow geometry of a reach (Benson and Dalrymple, 1967; Davidian, 1984), the errors in the slope-area calculations result mainly from the errors in measuring water-surface fallswd, Manning's n, expansion/contraction coefficients, and scour/fill relationships (Jarrett, 1985; Kirby, 1985). The best paleoflood sites are located in channel reaches with flow boundaries constrained by bedrock or other resistant boundary materials. Such channels produce relatively large stage changes for changes in flood discharge (Baker, 1977, 1984). Moreover, they do not change their cross sections appreciably during major floods. High levels of accuracy are difficult to obtain for high-gradient streams (Jarrett, 1984), expanding reaches, and especially for alluvial streams with a high scour uncertainty. Such reaches should be avoided, if possible.

Discharge estimation error in paleoflood studies is governed by the same factors as errors in other indirect flood measurements. However, because good paleoflood information tends to be preserved primarily at exceptionally stable geological sites, some of the principal error sources may be better controlled in paleoflood studies than in some historic or systematic flood studies. Methods of reducing paleodischarge estimation error in the various categories are discussed by Kochel et al. (1982) and Baker (1985).

Collection and Interpretation of Paleostage Data Paleostage information is obtained by determining the elevations of the upper surfaces of slackwater deposits. Conventional surveying methods (Benson and Dalrymple, 1967) are appropriate for determining the elevations, but specialized techniques are needed to identify the slackwater deposits and to establish the relationships between the deposits and the corresponding paleoflood water surfaces.

Assuming available bed material for transport, local sites of slackwater deposition may include mouths of tributaries, caves and rockshelters on canyon walls, flow-separation zones in abrupt channel expansions, and eddy zones or ineffective flow areas associated with channel bends or valley-side alcoves, channel constrictions, or other flow obstructions. Tributary mouth sites are very common. They occur because relatively small tributaries debouch their peak flows before mainstem flood peaks. The mainstem flooding may then backflood the tributary up to a level nearly equivalent to the mainstem flood stage. Bedrock caves are less common, but they are especially valuable for the long-term preservation of flood slackwater sediments. The cave environment prevents subsequent rainwater and tributary flow erosive effects. Reduced biological activity preserves deposit stratigraphy. The dynamics of flood slackwater sedimentation is described by Baker et al. (1983), Kochel et al. (1982), and Patton et al. (1979).

Interpretation of slackwater deposits in terms of paleostages, the dating of the deposits, and the interpretation of flood chronologies require principles of stratigraphy and geochronology. Stratigraphy is the branch of geology that deals with the arrangement of sedimentary strata, especially in relation to chronologic order of sequence. This aspect of a paleoflood study requires analysis by a specialist trained in stratigraphic geology. Individual flood sedimentation units are recognized, discriminated, and correlated among several sites along a paleoflood study reach (Kochel et al., 1982). Radiocarbon dating is the most common geochronologic tool used to provide absolute dates of the individual paleoflood events. A radiocarbon date is derived from a geochemical laboratory determination of the remaining present-day activity of the radiogenic isotope ^{14}C in an appropriate organic material (Faure, 1986; Stuiver and Polach, 1977). Analyses can be accomplished on small quantities of charcoal, seeds, and other organics that are commonly intercalated with ancient flood deposits. Floods in the period 1950 to present can be dated essentially to the calendar year (Baker et al., 1985). Floods in the period 1650 to 1950 require supplemental dating by historical documentation, archaeology, dendrochronology, or other means. Floods in the period 10,000 to 350 years ago generally can be dated with typical accuracy for that period by conventional radiocarbon procedures.

SWD-PSI studies can yield varying amounts of information on the paleostages and ages associated with ancient flow events. The minimal amount of information is a single paleostage indicator and a single date on the event. A somewhat more informative case is a single, vertically stacked sequence of dated slackwater deposits. Each successive deposit implies the occurrence of a paleoflood stage higher than the previous deposit. Thus the recorded flood series is censored and the censoring threshold increases when exceedances occur. However, the exact magnitude of exceedance is unknown, since various depths of flood water above the threshold are capable of emplacing a deposit.

In practice it is rare that a SWD–PSI study will yield only a single indicator. More commonly, numerous vertically stacked deposits along a reach are identified, traced, and correlated through stratigraphic analysis (Kochel et al., 1982). Floods that fail to leave deposits at one site, because they fail to exceed the local threshold, may be preserved at other sites. Deposits of a given paleoflood are traced laterally to their highest levels as they thin upstream along a tributary. Additional checks on the maximum flood level are provided by scour lines that can be traced to dated deposits. The highest deposit or mark of a given flood defines the magnitude of that flood. Thus, greater accuracy is achieved (at the cost of additional field work) when the slackwater paleoflood study attempts to document as many sites as possible in an appropriate study reach.

Availability of Paleoflood Data Although numerous useful SWD–PSI studies have been performed (see previous citations), there as yet exists no comprehensive archive of paleoflood data. At present, paleoflood data would have to be developed by individual at-site investigations at almost any site where it might be desired. As further paleoflood work is completed, this situation may change. Experience with paleoflood investigations has led to a recognition of regional factors conducive to slackwater sediment emplacement and preservation and to long-term channel stability, which is necessary for accurate transformation of paleoflood stages to discharges. These criteria include: (1) confined canyons or gorges developed in resistant geologic materials; (2) adequate concentrations of sand, silt, and coarser materials in transport; and (3) channel beds not subject to aggradation or degradation. These criteria are met in numerous upland areas of the West, the Midwest (Edwards and Ozark plateaus), the Appalachians, and New England. Appropriate sites for a SWD–PSI study should be identified by reconnaissance prior to investment in a detailed investigation.

A limiting factor in the adoption of SWD–PSI methodology is its multidisciplinary complexity. SWD–PSI studies require expertise in geology and geomorphology as well as in hydrology and hydraulics. In addition, the proper statistical interpretation of paleoflood data in regional flood-frequency analysis may require considerable statistical expertise and ingenuity. Computerized statistical procedures and data-management facilities for integrating historical and paleoflood data with systematic gaged records are not available. Lack of familiarity with concepts and terminology have hindered the adoption by engineers of paleoflood methodologies. Nonetheless, the expense of such studies is minor in relation to planning costs for major high-risk projects such as nuclear power plants or large dams. At present these opportunities are largely being ignored. The potential contribution of paleoflood data to augment the systematic records used in regional flood-frequency analyses should not be overlooked. At a minimum, the physical evidence of large paleofloods can be considered to provide objective evidence of the past occurrence and potential for recurrence of larger floods than might have been documented in systematic or historic flood records. For critical projects the paleoflood data should at least be collected, appropriately weighed, and considered in the overall decision process leading to design.

ADDITIONAL CHARACTERISTICS OF FLOOD DATA

Accurate determination of flood magnitudes is only one aspect of the general problem of accuracy of flood data. A general definition of data accuracy must include the ability of the data to support (or contradict) hypotheses or conclusions based on the data. Accuracy of flood data, therefore, must include its usability for drawing conclusions about flood frequency. These considerations apply to all types of flood data: streamflow data, rainfall data, and paleoflood data; subsequent references to flood data are intended to apply to all three types. Flood-data attributes that govern this aspect of accuracy include sample size, spatial and temporal correlation, spatial and temporal homogeneity, and the nature of the sampling plan governing the data-collection program. These attributes must be considered in conceptualizing the probabilistic generating mechanisms and populations used to represent the rainfall and flood processes and in selecting the statistical methods used to analyze the data.

Sample Size

Extreme or extraordinary floods are those that have return periods in excess of about 100 years or annual exceedance probabilities (tail probabilities) less than about 10^{-2}. The problem addressed by this report thus is the use of hydrologic theory and data to estimate annual exceedance probabilities that are less than 10^{-2}. It is known, however, that a data sample of size n, in the absence of a priori distributional assumptions, can furnish information only about exceedance probabilities greater than approximately $1/n$. For hydrologic records, n hardly ever is as large as 100, and so a single hydrologic record *cannot* furnish any direct information about the extreme tails of the flood-frequency distribution from which it was drawn.

Of course, one may fit a distribution to a sample of size 20 and use it to estimate 100-year or 10,000-year floods. The user must recognize, however, that such a model is based on extrapolation and that it begs the question of the form of the tail.

An obvious way to investigate the form of the flood-frequency tail thus is to enlarge the sample size. The station-year method (Fuller, 1914; Linsley et al., 1958) attempts to do this by pooling suitably normalized records from a number of stations into a single large sample of size equal to the total number of station years of record. Normalizations that have been used include division by the drainage area, division by the at-site sample mean, division by the estimated at-site 5-year flood (Rowe et al., 1957), or subtraction of the mean and division by the standard deviation.

Under the assumption that the pooled sample is equivalent to a random sample from a single homogeneous population, return periods of the order of the total number of station years presumably could be investigated. However, spatial correlation between samples tends to significantly reduce the total number of station years, i.e., the validity of the station-year approach depends on the stations being far enough apart to be somewhat independent, yet close enough together to be in the same hydrologically homogeneous region. These conflicting requirements and the realities of hydrologic record lengths tend to limit maximum station-year network sizes to a few hundred to a few thousand station years, with corresponding limits on the return periods that can be investigated.

Spatial Correlation

It is common knowledge that weather, climate, and hydrologic conditions all exhibit some degree of spatial structure. This spatial structure can be expressed statistically as a cross correlation among the records in a station-year network. Matalas and Benson

(1961) found that a network of 164 stream-gaging stations in New England had an average cross correlation of 0.26 for annual floods. With this degree of interstation correlation, the sample mean based on the pooling of 164 stations has the same variance as the sample mean based on only 4 independent stations (Matalas and Benson, 1961). Stedinger (1983) has found less stringent but similar results for sample variances and skew coefficients. These findings are roughly consistent with the fact that the same five or six major storms caused the three or four highest flood peaks (totalling about 500-600 station years) at most stations in the New England network (Matalas and Benson, 1961). In effect, the network did not see 500 independent floods during these storms, but rather about 80 representations of each of about six major regional storms. [See also Dalrymple (1960, p. 26) and Benson (1962, p. 25 ff).] It must be acknowledged that New England is small and more subject to large regional-scale storms than other parts of the country. Thus the effects of spatial correlation may be stronger there than elsewhere. Nonetheless, it appears that use of a k-station network instead of a single site is not likely to yield a full k-fold extension of the return periods that can be studied or to yield as much as a $\sqrt{k}$-fold improvement in estimation errors. Thus, the effects of interstation correlation need to be investigated for any frequency-analysis approach that involves pooling of flood or rainfall records. In particular, the maximum return period (or minimum exceedance probability) for which the pooled sample contains information should be critically evaluated.

Investigation of these matters would benefit from improvements in the structure and organization of the data bases used to store and retrieve flood data. Existing data bases have been organized primarily for convenience in storing and retrieving time-series data at fixed gage locations. The spatial aspects of the data have not been given high priority. Studies of interstation correlations, storm structure and movement, geographic distribution of floods, and other spatial characteristics of flood data would benefit from an improved ability to access spatial arrays of flood data at selected instants of time.

Nonstationarity

In addition to pooling concurrent records, available sample sizes sometimes may be enlarged by temporal extension. Information from historical sources such as newspaper articles, diaries, and personal recollections of long-time residents often can be used to reconstruct chronologies of historic floods (Thomson et al., 1964). If the historical records are detailed enough and hydraulic conditions have been sufficiently stable, it may be possible to determine the magnitudes, as well as the dates, of the flood events (Sutcliffe, 1985); this has been a very important source of flood-frequency information in China (Luo, 1985) and in other places (Tennessee Valley Authority, 1961). Paleohydrologic flood information similarly may be developed by geological and hydraulic analyses of fluvial deposits and landforms, as described elsewhere in this chapter and by Baker (1985).

Whenever historical or paleohydrologic information is pooled with more recent observations, the comparability of the two types of data comes into question. Several questions must be asked. First, it must be determined whether the hydrologic and hydraulic mechanisms that generated the historic floods and paleofloods were comparable to the mechanisms that generated the gaged events. In addition, it may be asked whether the generating mechanisms of the present and past will be representative of the mechanisms that will generate floods during the time period for which the flood frequency is being assessed. The fact of climatic change is well accepted in principle, although the magnitude and direction of any current change may be subject to debate and although it may be impossible to distinguish between irregular nonstationarities and random or correlation-induced fluctuations in a long-term stationary record. Similarly, changes in land use and hydraulic conditions in rivers are well known. USGS peak-flow records even contain qualification codes to document such changes. Finally, in view of the importance of measurement error, it may be asked whether differences in measurement technique affect the usability of the pooled data.

All of these questions apply also to long systematic (gaged) records, although they apply with more force to historic and paleohydrologic records. If the records are nonstationary, then the statistical characteristics of the nonstationarity need to be modeled, estimated, and extrapolated into the future before the records can be used.

These questions all have been widely discussed in the hydrologic and geoscience literature without achievement of a consensus on practically usable guidelines. Thus, these questions have to be investigated and resolved for each specific case in which historic or paleoflood data are to be incorporated into flood-frequency analysis.

Statistical Sampling Conditions

Finally, a fundamental problem exists in the statistical interpretation of paleoflood, historic, and miscellaneous-site flood measurements. As was discussed above, systematic records are relatively easy to interpret statistically, but nonsystematic records are not easy to interpret because they pertain to particular unique events rather than to the predefined classes of events that are the subject of probability and statistics. Whether a record is systematic or nonsystematic depends upon whether or not it was collected and preserved under a protocol that ensures that the record contains all occurrences of the events of interest and that it excludes all events that do not belong to the class of interest.

Records that are historic in the sense used here are by definition nonsystematic. Some records collected in the distant historic past, such as the records of the annual floods of the Nile at Alexandria, may in fact be systematic, despite their antiquity, provided that they were collected and preserved independently of the magnitudes of the events. It is conceivable that some of the historical texts pertaining to floods in China in fact may be systematic records of data such as tax receipts, agricultural yields, and expenses that are correlated with the occurrence of floods.

Much recent work on the treatment of historical information in flood-frequency analysis (Hirsh, 1985; Hydrology Subcommittee, 1982; Stedinger and Cohn, 1985, 1986a,b; Zhang, 1982) rests on assumptions that the historical records contain all occurrences of peaks exceeding some threshold. Such a record, in the terminology of Stedinger and Cohn (1985) and Hirsch (1985), is a censored sample and is in fact a special kind of systematic sample. The recording and preservation of historic information does not necessarily produce a record of this kind. It seems quite possible, for example, that of three floods ranked 3, 2, 1 and occurring in years 1850, 1875, and 1877, the 1875 flood might be superseded in the historical record by the 1877 flood, whereas the 1850 flood, by virtue of its primacy and isolation in time, might be remembered. Thus, another possible model for the record is that provided by Hosking and Wallis (1986a,b), who characterize the historical record in terms of knowledge only of the historical maximum event. The mechanisms by which historic flood events were detected, recorded, and preserved at a particular site were discussed in detail by Gerard and Karpuk (1979); further investigation, understanding, and mathematical modeling of these mechanisms is needed for reliable interpretation of historic records.

Similar remarks apply to paleohydrologic records. In this case, however, the mechanisms governing the recording and preservation of the events are geological and hydraulic. These physically based mechanisms may be somewhat more amenable to statistical modeling and interpretation than historic-record mechanisms.

Proper statistical treatment of historical and paleoflood data requires the use of auxiliary information in addition to the flood magnitudes. Such information includes the historical censoring threshold and historical period that enable a historical record to be treated as a censored sample. Some paleohistorical records may require multiple thresholds and historical periods or other as yet undetermined types of information for proper analysis. Existing flood-data archives have not been designed for storage and retrieval of the specialized parameters needed for proper interpretation and analysis of historical and paleoflood data. They also have not been designed for effective management and utilization of isolated flood determinations and their integration with systematic gaged records.

The importance of the statistical sampling plan is particularly apparent in the cases of the miscellaneous flood measurement records of the USGS and other agencies, and the U.S. Army Corps of Engineers' catalog of major storms. Both of these measurement programs were directed toward documentation of record-breaking events with the objective of providing information on extreme flood potentials, probable maximum precipitation (PMP), and probable maximum floods (PMF). Numerous significant storms and floods were never included in theses files simply because they did not significantly exceed previous events that already were recorded. These files thus may contain much useful data on magnitudes of record-breaking events, but they do not reflect the relative proportions of events of different magnitudes or the absolute rates of occurrence of events. Although these data sets tend to resemble sequences of record-breaking events, their modeling and interpretation are complicated by several factors. The essential problem is the spatial dimension of the data sets. In most cases only one event has been observed at any particular site. Thus the data sets contain few, if any, actual sequences of record-breaking events at the same site. But storm climatology varies from site to site, and flood magnitudes depend on site characteristics such as drainage area, slope, and soil type. Thus, it is not obvious how to pool observations made at different sites in a region to form sample sequences of record-breaking events usable for statistical analysis. The problem is further complicated because even some record-breaking events have been omitted from these files because they were superseded by still-larger events before they could be recorded in the files. Thus a concerted research effort would be required to develop a suitable methodology before meaningful statistical interpretations or estimates could be made from these files.

Seismologists have a somewhat similar problem in the analysis of earthquake records. Earthquakes, like storms and floods, occur "randomly" in time and space and have associated with them certain numerical measures that characterize their magnitudes or intensities. Similarly, there is a mixture of systematic and nonsystematic reporting: seismograph installations provide systematic records, whereas citizen reports of various kinds constitute nonsystematic records. The resulting earthquake records or catalogs tend to be quite complete for large recent earthquakes, but incomplete for small past earthquakes. The degree of completeness is related to the density and sensitivity of both the seismograph network and the general population. Stepp (1972) describes a method for accounting for this incompleteness by estimating earthquake occurrence rates separately for each earthquake intensity class using only the part of the catalog that is complete for that class. The period of completeness for each class was determined by plotting the estimated average arrival rate (number of occurrences divided by record length) as a function of record length, measured backward from the present time. The point at which the plot begins to decrease with record length is the beginning of the period of completeness. Veneziano and Van Dyck (1985) have generalized this model by considering spatial variability of the seismicity parameters and by characterizing the incompleteness by probabilities of detection by people and by instruments. The detection probabilities are correlated with population density and seismograph network density and sensitivity and thus are functions of space and time. A maximum likelihood technique is described for simultaneous estimation of all seismicity and incompleteness parameters.

It should be noted that the earthquake catalog completeness problem is not identical to the hydrometeorological one. For example, the earthquake catalog model does not address the fact that some major storms and floods simply have not been cataloged because they are not record-breaking events. Nonetheless, the probabilistic models and statistical estimation techniques used in earthquake catalog analyses are relevant and applicable in general terms to the hydrometeorological problem. Careful study and adaption of these techniques to the analysis of hydrometeorologic event catalogs would be well worthwhile.

6

Developmental Issues and Research Needs

As indicated throughout this report, the theoretical basis for the estimation of flood probabilities in the range of interest is in need of improvement. The report discusses opportunities to improve estimation through application of regional statistical analysis of streamflow, consideration of paleohydrologic data, and use of rainfall-runoff models driven by synthetically derived precipitation inputs. The committee recommends these new directions not because they are tested and proven, but because of their promise and apparent scientific cogency. There is little in this report that can be taken by the practitioner and applied without further development or research. However, with development (e.g., refinement of procedures and "how-to" guidance) and research, our ability to estimate probabilities of rare floods should improve. Specifically, the committee is confident that application of the recommendations of this report will be strengthened and more widely accepted through development and research on the following topics.

FLOOD STATISTICAL ANALYSIS

Regionalization techniques should be explored. Parametric model considerations have two principal components: spatial dependence and parametric structure. A study of joint distributions with prescribed marginal probabilities to account for spatial dependence is needed. Structure on the parameters among the sites is a key issue and a limited number of such structures have been studied to date. Existing index flood schemes imply a structure among the parameters, and the suitability of these implied structures needs attention. Explicit parametric structures ought to be based on studies of physical/hydrologic conditions and then validated. For nonparametric models, extrapolation techniques such as the Breiman-Stone (1985) methods should be extended to higher dimensions to learn how they behave in the presence of spatial dependence. Similarly, tail methods should be extended to higher dimensions.

RUNOFF MODELING

Each stage of a runoff modeling method for estimating exceedance probabilities of extreme floods involves methods which require research, testing, and development. The most critical issues involve the precipitation input. The synthetic storm approach, which involves estimation of probabilities of temporally and spatially averaged storm depths followed by reconstruction of temporal and spatial patterns for individual storm events, needs to be evaluated critically. Stochastic rainfall models could be used effectively to do this. The critical assumption of storm transposition should be tested; that is, along with storm catalog development the various steps of implementation of storm transposition need testing. The data enhancement proposed in the next section will provide a solid base for research. We strongly recommend research on stochastic rainfall models, particularly emphasizing upper tail behavior and regionalization. Storm transposition based on a historic storm catalog needs much additional research.

A second general issue for additional research in runoff modeling is the capability of existing rainfall-runoff transformation models to simulate extreme events accurately. This issue could be addressed through modeling experiments on historically observed large floods for which reliable data are available on both rainfall and streamflow. Such experiments would be useful for comparing models, estimating the likely magnitude of errors, and identifying physical mechanisms which may not be operating at lower discharges.

Several other topics require research, including the problem of estimating probability distributions for antecedent conditions, for snowmelt rates and for the occurrence of frozen ground. Methods for error analysis need to be developed, so that some kind of measure of uncertainty can be applied to probabilities estimated by runoff modeling techniques.

DATA CONSIDERATIONS

Most of the approaches discussed in this report have been inhibited in development and verification by a lack of systematic data. Major efforts are needed to compile comprehensive data bases for developing and testing approaches proposed in chapters 3 and 4. Until this is done, the level of confidence needed for field implementation of techniques for estimating probabilities of extreme floods will be severely limited.

The primary step in the statistical strategy described in chapter 3 was the compilation of a large set of flood records from sites that are hydrologically similar to the sites of interest. The availability of the peak flow records, as described in chapter 5, is on the order of one station per 150 square miles, and the average record length is 22 years. Regionalization of these data potentially adds to the effective sample length; however, it is questionable, even when regionalization is used, whether currently available records are sufficient to model 1,000- to 10,000-year return-period floods. The consideration of paleoflood and historical flood data and their incorporation into regional flood-frequency analyses are worthy of further development and research. Lack of familiarity with these concepts has hindered the gathering of such data, and there may be statistical difficulties in the consideration of paleoflood and historical data in regional analyses. Research is needed on the mechanisms by which paleoflood and historical flood records are produced and preserved with an emphasis on constructing statistical models that reflect these mechanisms, as well as consideration of climatological changes that may have occurred between the historical event and the contemporary data record.

Efforts also should be made to develop flood data storage and retrieval systems that can help to integrate isolated historical and paleoflood measurements and measurements at miscellaneous sites with systematic records collected at gaged sites. Such systems should support spatial representation of flood data as well as the traditional point-time-series representations. In addition, provision should be made for storing and retrieving the various types of auxiliary information needed for statistical analysis of historical and paleoflood data.

Runoff modeling approaches described in chapter 4 depend on

the adequacy and accuracy of available precipitation records. The use of continuous models requires continuous data (at least for the duration of the events that are simulated). Furthermore, many applications of space–time models have been considered infeasible because of the lack of sufficiently dense gage networks.

The primary data sets available for use with runoff models are the systematic hourly and daily precipitation data, and the storm catalog. However, the large size of the systematic hourly and daily precipitation data sets accessible at NCDC makes data management and quality control formidable tasks. Additionally, a high percentage of these data are unrelated to severe floods of interest. The storm catalog (upon which the storm transposition methods introduced in the section on synthetic storms in chapter 4 were based) was developed by the National Weather Service and U.S. Army Corps of Engineers. Methods using transposition, such as those proposed in this report, are based on the assumption that the input data include a complete and uniform record of "significant" storms. The storm catalog, as described, is a unique and essential data set. However, it does not form a systematic record and the quality of the storm estimates is not consistent from storm to storm.

Given these substantial data limitations, probably the most important of several data development efforts would involve understanding how the systematic NCDC precipitation records could be used to complete the storm catalog. Relationships should be developed between storms described in both the storm catalog and the continuous (digital) rainfall records. From such comparisons, adjustment factors may be determined to adjust storm events recorded in the archived data set but not in the storm catalog, so that such storms could be represented in the storm catalog, making the catalog comprehensive for storms with accumulations above some specified significance level. Assessments of the storm events described by the archived data (the systematic NCDC records) will benefit by comparison with assessments of the same events described in the storm catalog, because of the greater density of observations associated with bucket surveys (used in analyzing cataloged storms). This process will improve the assessment of storm means and total volumes. The resulting storm catalog should be structured for easy access by scientists and engineers working on probabilities of hydrologic catastrophes.

Other important data development projects are: to further automate and speed processing of storms not yet included in the catalog;

to find ways to simplify the procedures (possibly using network design principles) or reduce the cost and effort involved in reviewing all of the systematic rainfall records; and to expand the understanding of homogeneous regions and regionalization of hydrologic events.

A COMPREHENSIVE STATISTICAL MODEL

A comprehensive statistical model needs to be developed for relating probability estimates based on flood statistics to estimates based on runoff modeling. One possibility is to use the stochastic structure of the storm-rainfall and antecedent-condition process to induce a stochastic structure (derived distribution) on the flood process by means of a deterministic rainfall-runoff transformation (Eagleson, 1972). This is discussed in chapter 4. Also, as discussed in Chapter 5, combined statistical-deterministic models have become highly developed in the area of seismic risk analysis (Veneziano and Van Dyck, 1985). Many valuable lessons might be learned from study of this work. In addition to whatever theoretical insights the derived distribution might furnish about flood tail probabilities, an integrated statistical-deterministic hydrologic model would permit joint use of all available rainfall and flood-flow data in estimation of extreme flood probabilities.

References

AGU Section on Hydrology. 1965. Inventory of Representative and Experimental Watershed Studies Conducted in the United States. American Geophysical Union, Washington, D.C.

Alexander, G. N. 1963. Using the probability of storm transposition for estimating the frequency of rare floods. Journal of Hydrology 1:46–57.

Alexander, G. N. 1969. Application of probability to spillway design flood estimation. In: Proceedings of the Leningrad Symposium on Floods and Their Computation, August 1967, Vol. 1, pp. 536–543. International Association of Scientific Hydrology, Gentbrugge, Belgium.

Baker, V. R. 1974. Paleohydraulic interpretation of quaternary alluvium near Golden, Colorado. Quaternary Research 4:95–112.

Baker, V. R. 1977. Stream-channel response to floods with examples from central Texas. Geological Society of America Bulletin 88:1057–1071.

Baker, V. R. 1983. Paleoflood Hydrologic Techniques for the Extension of Streamflow Records. Transportation Research Board, Publ. No. 922, pp. 18–23. National Research Council, Washington, D.C.

Baker, V. R. 1984. Flood sedimentation in bedrock boundary fluvial systems. In: E. H. Koster and R. J. Steel (eds.), Sedimentology of Gravels and Conglomerates, pp. 87–98. Canadian Society of Petroleum Geologists Memoir No. 10, Calgary, Alberta.

Baker, V. R. 1985. Paleoflood hydrology and extraordinary flood events. Paper presented at U.S.-China Bilateral Symposium on the Analysis of Extraordinary Flood Events, Nanjing, China. Journal of Hydrology 96(1–4):79–99. In press.

Baker, V. R., R. C. Kochel, and P. C. Patton. 1979. Long-Term Flood Frequency Analysis Using Geological Data. International Association of Hydrological Sciences Publ. No. 128, pp. 3–9.

Baker, V. R., R. C. Kochel, P. C. Patton, and G. Pickup. 1983. Palaeohydrologic analysis of Holocene flood slack–water sediments. In: J. Collinson and J. Lewin (ed.), Modern and Ancient Fluvial Systems, pp. 229–239. International Association of Sedimentologists Spec. Publ. No. 6.

Baker, V. R., G. Pickup, and H. A. Polach. 1985. Radiocarbon dating of flood events, Katherine Gorge, Northern Territory, Australia. Geology 13:344–347.

Barnes, H. H., Jr., and Jacob Davidian. 1978. Indirect methods. In: R. W. Herschy (ed.), Hydrometry: Principles and Practice, Ch. 5, pp. 149–204. John Wiley, New York.

Battan, L. J. 1973. Radar Observations of the Atmosphere, rev. ed. University of Chicago Press, Chicago, Illinois. 324 pp.

Benson, M. A. 1962. Evolution of Methods for Evaluating the Occurrence of Floods. U.S. Geological Survey Water Supply Paper No. 1580-A. 30 pp.

Benson, M. A. and T. Dalrymple. 1967. General field and office procedures for indirect discharge measurements. In: Techniques of Water-Resource Investigations, Book 3, Ch. A1. U.S. Geological Survey. 30 pp.

Beven, K. J. 1975. Towards the use of catchment geomorphology in flood frequency predictions. In: Earth Surface Processes and Landforms.

Beven, K. J., and Hornberger. 1982. Assessing the effects of spatial pattern of precipitation in modeling streamflow hydrographs. Water Resources Bulletin 18(6):823–829.

Boos, D. 1984. Using extreme value theory to estimate large percentiles. Technometrics 26(1):33–39.

Brakensiek, D. L., H. B. Osborn, and W. J. Rawls. 1979. Field Manual for Research in Agricultural Hydrology. Agricultural Handbook No. 224. U.S. Department of Agriculture.

Bras, R. L., D. R. Gaboury, D. S. Grossman, and G. J. Vicens. 1985. Spatially varying rainfall and flood risk analysis. Journal of Hydraulic Engineering 111(5):754–773.

Breiman, L. and C. J. Stone. 1985. Broad Spectrum Estimates and Confidence Intervals for Tail Quantiles. Tech. Rep. No. 46. Department of Statistics, University of California, Berkeley.

Bullard, K. L. 1986. Comparison of Estimated Probable Maximum Flood Peaks with Historic Floods. Hydrology Branch, Engineering and Research Center, Bureau of Reclamation, Denver. 165 pp.

California Department of Water Resources. Series. Water Conditions in California. Snow Survey Cooperators, Department of Water Resources, Sacramento, Calif.

Chow, V. T. 1964. Handbook of Applied Hydrology. McGraw Hill, New York. 1,467 pp.

Church, M. 1978. Palaeohydrological reconstructions from a Holocene valley fill. In: A. D. Miall (ed.), Fluvial Sedimentology, pp. 743–772. Canadian Society of Petroleum Geologists Mem. No. 5, Calgary, Alberta.

Cook, J. L. 1985. Quantifying peak discharges for historical floods. Paper presented at United States–China Bilateral Symposium on the Analysis of Extraordinary Flood Events, Nanjing, China. Journal of Hydrology 96(1–4):29–40. In press.

Costa, J. E. 1978. Holocene stratigraphy in flood-frequency analysis. Water Resources Research 14:626–632.

REFERENCES

Costa, J. E. 1983. Paleohydraulic reconstruction of flash-flood peaks from boulder deposits in the Colorado Front Range. Geological Society of America Bulletin 94:986–1004.

Costa, J. E. 1985. Interpretation of the largest rainfall-runoff floods measured by indirect methods on small drainage basins in the conterminous United States. Paper presented at United States–China Bilateral Symposium on the Analysis of Extraordinary Flood Events, Nanjing, China. Journal of Hydrology 96(1–4):101–115. In press.

Costa, J. E., and V. R. Baker. 1981. Surficial Geology. John Wiley & Sons, New York. 498 pp.

Crawford, K. 1979. Considerations for the design of a hydrologic data network using multivariate sensors. Water Resources Research 15(6):1752–1761.

Creager, W. P., and J. D. Justin. 1950. Hydroelectric Handbook. John Wiley & Sons, New York. 66 pp.

Crippen, J. R., and C. D. Bue. 1977. Maximum Flood Flows in the Conterminous United States. U.S. Geological Survey Water Supply Paper No. 1887. 52 pp.

Dalrymple, Tate. 1960. Flood-Frequency Analyses. U.S. Geological Survey Water Supply Paper No. 1543-A. 80 pp.

Dalrymple, Tate, and M. A. Benson. 1967. Measurement of peak discharge by the slope-area method. U.S. Geological Survey Techniques of Water Resources Investigations, Book 3, Ch. A2. 12 pp.

Davidian, J. U.S. Geological Survey. 1984. Computation of Water-Surface Profiles in Open Channels. Techniques of Water-Resources Investigations, Book 3, Ch. A15. 48 pp.

Davis, R., and S. Resnick. 1984. Tail estimates motivated by extreme value theory. Annals of Statistics 12:1467–1487.

Dawdy, D. R., R. W. Lichty, and J. J. Bergmann. 1972. A Rainfall-Runoff Simulation Model for Estimation of Flood Peaks for Small Drainage Basins. U.S. Geological Survey Prof. Paper No. 506-B. 28 pp.

Dury, G. H. 1976. Discharge prediction, present and former, from channel dimensions. Journal of Hydrology 30:219–245.

Eagleson, P. S. 1972. Dynamics of flood frequency. Water Resources Research 8(4):878–898.

Efron, Bradley. 1982. The Jackknife, the Bootstrap, and Other Resampling Plans. Society of Industrial and Applied Mathematics, Philadelphia. 92 pp.

Ely, L. L., and V. R. Baker. 1985. Reconstructing paleoflood hydrology with slackwater deposits: Verde River, Arizona. Physical Geography 6(2):103–126.

Ethridge, F. G., and S. A. Schumm. 1978. Reconstructing paleochannel morphologic and flow characteristics: limitations and assessment. In: A. D. Miall (ed.), Fluvial Sedimentology, pp. 703–721. Canadian Society of Petroleum Geologists Mem. No. 5, Calgary, Alberta.

Environmental Data Service. 1968. Climatic Atlas of the United States. Environmental Science Services Administration, U.S. Department of Commerce, Washington, D.C. 80 pp.

Farnsworth, R. K., and E. S. Thompson. 1982. Mean Monthly, Seasonal, and Annual Pan Evaporation for the United States. NOAA Tech. Rep. NWS 34. U.S. Department of Commerce, Washington, D.C. 82 pp.

Farnsworth, R. K., E. S. Thompson, and E. L. Peck. 1982. Evaporation Atlas for the Contiguous 48 United States. NOAA Tech. Rep. NWS 33. U.S. Department of Commerce, Washington, D.C. Four maps, 26 pp.

Farnsworth, R. K., E. C. Barrett, and M. S. Dhanju. 1984. Application of Remote Sensing to Hydrology Including Ground Water. Technical Documents in Hydrology. International Hydrological Program, UNESCO, Paris, France. 122 pp.

Faure, G. 1986. Principles of Isotope Geology. John Wiley, New York. 592 pp.

Feldman, A. D. 1981. HEC models for water resources system simulation, theory and experience. In: V. T. Chow (ed.), Advances in Hydroscience, Vol. 12, pp. 297–423. Academic Press, New York.

Foley, M. G., J. M. Doesburg, and D. A. Zimmerman. 1984. Paleohydrologic techniques with environmental applications for siting hazardous waste facilities. In: E. H. Koster and R. J. Steel (ed.), Sedimentology of Gravels and Conglomerates, pp. 99–108. Canadian Society of Petroleum Geologists Mem. No. 10, Calgary, Alberta.

Franz, D., B. Kraeger, and R. Linsley. 1986. A System for Generating Long Streamflow Records for Study of Floods of Long Return Period, for presentation at the International Symposium on Flood Frequency and Risk Analysis.

Frederick, R. H., V. A. Myers, and E. O. Auciello. 1977. Five to 60 Minute Precipitation Frequency for the Eastern and Central United States. NOAA Tech. Memo NWS HYDRO 35. U.S. Department of Commerce, Silver Spring, Md.

Fuller, W. E. 1914. Flood flows. Transactions of ASCE 77:564–617.

Gerard, R., and E. W. Karpuk. 1979. Probability analysis of historical flood data. Journal of Hydraulics Division, American Society of Civil Engineers 105(HY-9):1153–1165.

Gregory, K. J. (ed.). 1983. Background to Palaeohydrology: A Perspective. John Wiley & Sons, New York. 486 pp.

Greene, D. R., J. D. Nilsen, R. E. Saffle, D. W. Holmes, M. D. Hudlow, P. R. Ahnert. 1983. RADAP II: An Interim Radar Data Processor. Preprints, 21st Conference on Radar Meteorology, pp. 404–408. AMS and Alberta Research Council, Canadian Meteorological and Oceanographic Society, Edmonton, Alberta, Canada.

Hatch, W. L. 1983. Selective Guide to Climatic Data Sources. Key to Meteorological Records Documentation No. 411. National Climatic Data Center, NOAA, U.S. Department of Commerce, Asheville, N.C. 338 pp.

Hauth, L. D. 1974. Model Synthesis in Frequency Analysis of Missouri Floods. U.S. Geological Survey Circ. 708. 16 pp.

Hazen, Allen. 1930. Flood Flows. Wiley, New York, 199 pp.

Hershfield, D. M. 1961. Rainfall–Frequency Atlas of the United States for Durations from 30 Minutes to 24 Hours and Return Periods from 1 to 100 Years. U.S. Weather Bureau Tech. Pap. 40. U.S. Department of Commerce, Washington, D.C. (out of print).

Hirsch, R. M. 1985. Probability plotting position formulas for flood records with historical information. Paper presented at United States–China Bilateral Symposium on the Analysis of Extraordinary Flood Events, Nanjing, China. Journal of Hydrology 96(1–4):185–199. In press.

REFERENCES

Hosking, J. R. M. 1986a. The Theory of Probability Weighted Moments. BIM Research Report.
Hosking, J. R. M. 1986b. The Wakeby Distribution. IMB Research Report.
Hosking, J. R. M., J. Wallis, and E. Wood. 1985a. Estimation of generalized extreme value distribution by the method of probability-weighted moments. Technometrics 27(3)231-262.
Hosking, J. R. M., J. Wallis, and E. Wood. 1985b. An appraisal of the regional flood frequency procedure in the UK flood studies report. Journal of Hydrological Sciences 30(1)85-109.
Hosking, J. R. M., and J. R. Wallis. 1986a. Paleoflood hydrology and flood frequency analysis. Water Resources Research 22(4):543-550.
Hosking, J. R. M., and J. R. Wallis. 1986b. The value of historical data in flood frequency analysis. Water Resources Research 22(11):1606-1612.
Hosking, J. R. M., and J. R. Wallis. 1987. An index flood procedure for regional rainfall frequency analysis. Presented at American Geophysical Union, Baltimore, Md.
Hutchison, N. E. (compiler). 1975. Watstore User's Guide, Vol. 1. U.S. Geological Survey Open File Report 75-426.
Hydrologic Research Laboratory Staff. 1972. National Weather Service River Forecast System, Forecast Procedures. NOAA Technical Memo NWS HYDRO-14, U.S. Department of Commerce, Silver Spring, Md.
Hydrology Subcommittee. 1982. Guidelines for Determining Flood Flow Frequency. Hydrology Subcommittee Bulletin 17-B, with editorial corrections. Interagency Advisory Committee on Water Data, Office of Water Data Coordination, U.S. Geological Survey, 28 pp.
Hydrology Subcommittee. 1986. Feasibility of Assigning a Probability to the Probable Maximum Flood. Interagency Advisory Committee on Water Data, Office of Water Data Coordination, U.S. Geological Survey. 79 pp.
Jarrett, R. D. 1984. Hydraulics of high gradient streams. Journal of Hydraulics Division, American Society of Civil Engineers 110(11):1519-1539.
Jarrett, R. D. 1985. Errors in slope-area computations of peak discharges in mountain streams. Paper presented at U.S.-China Bilateral Symposium on the Analysis of Extraordinary Flood Events, Nanjing, China. Journal of Hydrology. In press.
Jarvis, C. S. 1926. Flood flow characteristics. Transactions of the American Society of Civil Engineers 89:985-1032.
Kavvas, M. L., and K. R. Herd. 1985. A radar-based stochastic model for short-time-increment rainfall. Water Resources Research 21(9):1437-1455.
Kirby, W. 1974. Algebraic boundedness of sample statistics. Water Resources Research 10(2):220-222.
Kirby, W. 1985. Statistical error analysis of the slope-area method of indirect discharge determination. Paper presented at United States-China Bilateral Symposium on the Analysis of Extraordinary Flood Events, Nanjing, China. Journal of Hydrology 96(1-4):125-138. In press.
Kochel, R. C., and V. R. Baker. 1982. Paleoflood hydrology. Science 215:353-361.
Kochel, R. C., V. R. Baker, and P. C. Patton. 1982. Paleohydrology of Southwestern Texas. Water Resources Research 18:1165-1183.

Krajewski, W. F., and M. D. Hudlow. 1983. Evaluation and Application of a Real-Time Method to Estimate Mean Areal Precipitation from Rain Gage and Radar Data. In: Proceedings, Technical Conference on Mitigation of Natural Hazards Through Real-Time Data Collection Systems and Hydrological Forecasting, pp. 18–19. WMO, State of California Department of Water Resources, and NOAA, Sacramento, Calif.

Krug, W. R., and G. L. Goddard. 1985. Effects of Urbanization on Streamflows, Sediment Loads, and Channel Morphology in Pheasant Branch Basin near Middletown Wisconsin. U.S. Geological Survey, Water Resources Investigation Report 85-4068.

Larson, L. W., and E. L. Peck. 1974. Accuracy of precipitation measurements for hydrologic modeling. Water Resources Research 10(4):857–863.

Leadbetter, M. R., G. Lindgren, and H. Rootzen. 1983. Extremes and Related Properties of Random Sequences and Processes. Springer-Verlag.

Lepkin, W. D., M. McDeLapp, W. H. Kirby, and T. A. Wilson. 1979. Watstore User's Guide, Vol. 4. U.S. Geological Survey Open-File Report 79-1336-I.

Lettenmaier, D. P. and K. Potter. 1985. Testing flood frequency estimation methods using a regional flood generation model. Water Resources Research 21(12)1903–1914.

Lettenmaier, D. P., Wallis, J. R., and Wood E. 1986. Effect of regional heterogeneity on flood-frequency estimation. Water Resources Research.

Linsley, R. K., Jr., Kohler, M. A., and Paulhus, J. L. H. 1958. Hydrology for Engineers. McGraw-Hill, New York, 340 pp.

Linsley, R. K., Jr., Kohler, M. A., and Paulhus, J. L. H. 1982. Hydrology for Engineers, 3rd ed. McGraw-Hill, New York. 496 pp.

Ludlum, D. M. 1962. Extremes for snowfall in the United States. Weatherwise 15(6):246–253.

Luo Cheng-Zheng. 1985. Investigation and regionalization of historical floods in China. Paper presented at United States–China Bilateral Symposium on the Analysis of Extraordinary Flood Events, Nanjing, China. Journal of Hydrology 96(1–4):41–51. In press.

Madsen, H. O., S. Krenk, and N. C. Lind. 1986. Methods of Structural Safety. Prentice-Hall, Englewood Cliffs, N.J. 403 pp.

Marino, L., and A. Bradley, Jr. 1986. Precipitation frequency-runoff frequency relationship in hydrologic design (abst.). EOS, Transactions, American Geophysical Union 67(44):933.

Matalas, N. C. and M. A. Benson. 1961. Effect of interstation correlation on regression analysis. Journal of Geophysical Research 66(10):3285–3293.

Miller, J. R., and R. H. Frederick. 1966. Normal Monthly Number of Days with Precipitation of 0.5, 1.0, 2.0, and 4.0 Inches or More in Conterminous United States. ESSA Technical Paper 57. U.S. Department of Commerce, Washington, D.C.

Miller, J. R., R. H. Frederick, and R. J. Tracy. 1973. Precipitation-Frequency Atlas of Western United States (by state). NOAA Atlas 2, 11 vol., U.S. Department of Commerce, Silver Spring, Md.

Myers V. A. 1970. Joint Probability Method of Tide Frequency Analysis Applied to Atlantic City and Long Beach Island, N.J. ESSA Technical Memorandum WBTM HYDRO 11. Environmental Science Services Administration, Silver Spring, Md.

National Research Council. 1983a. Evaluation of the FEMA Model for Estimating Potential Coastal Flooding from Hurricanes and Its Application to Lee County, Florida. National Academy Press, Washington, D.C.

National Research Council. 1983b. Safety of Existing Dams: Evaluation and Improvement. National Academy Press, Washington, D.C.

National Research Council. 1985. Safety of Dams: Flood and Earthquake Criteria. National Academy Press, Washington, D.C.

Newton and Cripe. 1973. Flood studies for safety of TVA nuclear plants—hydrologic and embankment breaching analysis. Paper presented at the 1973 ASCE National Water Resources Engineering Meeting, Washington, D.C.

O'Connor, J. E., R. H. Webb, and V. R. Baker. 1986. Paleohydrology of pool and riffle pattern development, Boulder Creek, Utah. Geological Society of America Bulletin 97(4):410–420.

Office of Water Data Coordination. 1977. National Handbook of Recommended Methods for Water Data Acquisition. U.S. Geological Survey, Reston, Va.

Partridge, J., and V. R. Baker. 1987. Paleoflood hydrology of the Salt River, Central Arizona. Earth Surface Processes and Landforms 12:109–125.

Patton, P. C., V. R. Baker, and R. C. Kochel. 1979. Slackwater deposits: a geomorphic technique for the interpretation of fluvial paleohydrology. In: D. P. Rhodes and G. P. Williams (ed.), Adjustments of the Fluvial Systems, pp. 225–253. Kendal/Hunt Publishing Company, Dubuque, Iowa.

Paulhus, J. L. E., and J. F. Miller. 1957. Flood frequencies derived from rainfall data. Paper No. 1451, ASCE Journal of Hydraulics Division, pp. 1–18.

Peck, E. L. 1972a. Snow measurement predicament. Water Resources Research 8(1):244–248.

Peck, E. L. 1972b. Review of Methods of Measuring Snow Cover, Snowmelt, and Streamflow under Winter Conditions. In: Proceedings, International Symposium on the Role of Snow and Ice in Hydrology. Banff, Canada, pp. 582–597. IAHS-AISH Publ. No. 107, Vol. 1.

Pickands, J. 1975. Statistical inference using extreme order statistics. Annals of Statistics 3:119–130.

Potter, K. W., and J. F. Walker. 1982. Modelling the error in flood discharge measurements. In: A. H. El-Shaarawi (ed.), Time Series Methods in Hydrosciences, Vol. 17 of Developments in Water Science. Elsevier, Amsterdam.

Prescott, P., and A. Walden. 1980. Maximum likelihood estimation of the parameters of the generalized extreme value distribution. Biometrika 67:723–724.

Prescott, P., and A. Walden. 1983. Maximum likelihood estimation of the three-parameter generalized extreme value distribution. J. Statist. Comput. Simul. 16:241–250.

Pugsley, W. I. (ed.). 1981. Flood Hydrology Guide for Canada: Hydrometeorological Design Techniques. Report No. CLI 3-81. Environment Canada, Atmospheric Environment Service, Downsview, Ontario. 102 pp.

Rantz, S. E. 1982. Measurement and Computation of Streamflow, 2 vols. U.S. Geological Survey Water Supply Paper 2175. 631 pp.

Richards, F. P., and R. G. Wescott. 1986. Very low probability precipitation-frequency estimates—a perspective. Presented at International Symposium on Flood Frequency and Risk Analysis, Baton Rouge, La., May 14–17, 1986.

Riedel, J. T., and L. C. Schreiner. 1980. Comparison of Generalized Estimates of Probable Maximum Precipitation with Greatest Observed Rainfalls. NOAA Technical Report NWS 25, Washington, D.C.

Rodriguez-Iturbe, I., and J. R. Valdes. 1979. The geomorphic structure of hydrologic response. Water Resources Research 15(6):1409–1420.

Rowe, R. R., G. L. Long, and T. C. Royce. 1957. Flood frequency by regional synthesis. Transactions of the American Geophysical Union 38(6):879–884.

Schaefer, M. C. 1987. Analysis of precipitation data for the state of Washington. Paper present at the AGU National Meeting, San Francisco, December 1987.

Schilling, W., and L. Fuchs. 1986. Errors in stormwater modeling—a quantitative assessment. Journal of Hydraulic Engineering 112(2):111–123.

SCS–USDA. Series. Water Supply Outlook and Federal–State–Private Cooperative Snow Surveys. Soil Conservation Service, West Technical Service Center, Portland, Oreg.

Sevruk, B. 1982. Methods of Correction for Systematic Error in Point Precipitation Measure for Operational Use. Operational Hydrology Report No. 21, World Meteorological Organization No. 589. Geneva, Switzerland. 91 pp.

Shipe, A. P., and J. T. Riedel. 1976. Greatest Known Areal Storm Rainfall Depths for the Contiguous United States. NOAA Tech. Memo. NWS HYDRO-33. 174 pp.

Smith, J. A. 1986. Estimating the Upper Tail of Flood Frequency Distribution. Preprint.

Smith, R. L. 1986. Threshold models for sample extremes. In: J. Tiago de Oliveira (ed.), Statistical Extremes and Applications, pp. 621–638. Reidel, Dordrecht.

Statistical Science. February 1987. The Calculus and Uncertainty in Artificial Intelligence and Expert Systems, Vol. 2, No. 1.

Stedinger, J. R. 1983. Estimating a regional flood frequency distribution. Water Resources Research 19(2):503–510.

Stedinger, J. R. and T. A. Cohn. 1985. The use of historical information in flood frequency analysis. Paper presented at United States–China Bilateral Symposium on the Analysis of Extraordinary Flood Events, Nanjing, China. Journal of Hydrology 96(1–4):215–223. In press.

Stedinger, J. R., and T. A. Cohn. 1986a. The value of historical and paleoflood information in flood frequency analysis. Water Resources Research 22(5):785–793.

Stedinger, J. R., and T. A. Cohn. 1986b. Historical flood frequency: Its value and use. Presentation at the International Symposium on Flood Frequency and Risk Analysis, Louisiana State University, Baton Rouge, La.

Stepp, J. C. 1972. Analysis of completeness of the earthquake sample in the Puget Sound area and its effect on statistical estimates of earthquake hazard. In: Proceedings of a International Conference on Microzonation, Vol. 2, pp. 897–910.

Stuiver, M., and H. A. Polach. 1977. Discussion—Reporting of ^{14}C data. Radiocarbon 19:355–363.

Sutcliffe, J. V. 1985. The use of historical records in flood frequency analysis. Paper presented at United States–China Bilateral Symposium on the Analysis of Extraordinary Flood Events, Nanjing, China. Journal of Hydrology 96(1–4):159–171. In press.

REFERENCES

Tennessee Valley Authority. 1961. Floods and Flood Control. Technical Report No. 26, Knoxville, Tenn.

Thom, H. C. S. 1957. Probabilities of one-inch snowfall thresholds for the United States. Monthly Weather Review 85(8):269–271.

Thomas, W. O., Jr. 1982. An evaluation of flood frequency estimates based on rainfall/runoff modeling. Water Resources Bulletin 18(2):221–230.

Thomas, W. O., Jr. 1986a. Comparison of flood-frequency estimates based on observed and model-generated peak flows. Proceedings of the International Symposium on Flood Frequency and Risk Analysis, Louisiana State University, Baton Rouge, La.

Thomas, W. D., Jr. 1986b. The role of flood-frequency analysis in the U.S. Geological Survey. In: Proceedings of the International Symposium on Flood Frequency and Risk Analysis, Louisanna State University, Baton Rouge, La.

Thomson, M. T., W. B. Gannon, M. P. Thomas, and G. S. Hayes. 1964. Historical floods in New England. U.S. Geological Survey Water Supply Paper 1779-M. 105 pp.

Thurman, J. L., and R. T. Roberts. 1987. Hydrologic Data for Experimental Agricultural Watersheds in the United States, 1977. U.S. Department Agriculture Misc. Publ. 1454. 341 pp.

Troutman, B. M. 1983. An analysis of input errors in precipitation-runoff models using regression with errors in the independent variables. Water Resources Research 19(4):947–964.

Tucker, L. S. 1969. Raingage Networks in the Largest Cities. ASCE Urban Water Resources Research Program Technical Memo No. 9. 90 pp.

Tucker, L. S. 1970a. Non-Metropolitan Dense Raingage Networks. ASCE Urban Water Resources Research Program Technical Memo No. 11. 51 pp.

Tucker, L. S. 1970b. Availability of Rainfall-Runoff Data for Partly Sewered Urban Drainage Catchments. ASCE Urban Water Resources Research Program Technical Memo No. 13. 156 pp.

U. S. Army Corps of Engineers. 1945 to present. Storm Rainfall in the United States (ongoing publication of major storm data). Office of the Chief of Engineers, Washington, D.C.

Valdes, J. B., I. Rodriguez-Iturbe, and V. K. Gupta. 1985. Approximations of temporal rainfall from a multidimensional model. Water Resources Research 21(8):1259–1270.

Veneziano, D., and J. Van Dyck. 1985. Draft report. Analysis of earthquake catalogs for incompleteness and recurrence rates. Appendix A-6 in Seismic Hazard Methodology for Nuclear Facilities in the Eastern United States. Electric Power Research Institute.

Wallis, J. R., N. C. Matalas, and J. R. Slack. 1974. Just a moment. Water Resources Research 10(2).

Wallis, J. R. 1982. Probable and Improbable Rainfall in California. IBM Research Report RC 9350, pp. 1–26.

Wallis, J. R., and E. F. Wood. 1985. Relative accuracy of Log Pearson III procedures. Journal of Hydraulic Engineering 111(7):1043–1056.

Waymire, E., and V. K. Gupta. 1981. The mathematical structure of rainfall representations, 1. A review of the stochastic rainfall models. Water Resources Research 17(5):1261–1272.

Waymire, E., V. K. Gupta, and I. Rodriguez-Iturbe. 1984. A spectral theory of rainfall intensity at the Meso-Beta scale. Water Resources Research 20(10):1453–1465.

WB-ESSA, 1970. Weather Bureau Observing Handbook No. 2: Substation Observations. Office of Meteorological Operations, Weather Bureau, Environmental Science Services Administration, Silver Spring, Md. 77 pp.

Weather Bureau. 1946. Manual for Depth–Area–Duration Analysis of Storm Precipitation. Cooperative Studies Technical Report No. 1. U.S. Departments of Commerce and Interior, Washington, D.C. 69 pp.

Weather Bureau. 1958. Rainfall–Intensity–Frequency Regime: Southeastern United States. Technical Paper No. 29, Part 2. U.S. Department of Commerce, Washington, D.C.

Weissman, I. 1978. Estimation of parameters and large quantiles based on the K largest observation. Journal of the American Statistical Association 73:812–815.

Wenzel, H. G., Jr. 1982. Rainfall for urban stormwater design. In D. F. Kibler (ed.), Urban Stormwater Hydrology. Water Resources Monograph 7. American Geophysical Union, Washington, D.C.

Wilkinson, J. C., and L. V. Tavares. 1972. A methodology for the synthesis of spatially-distributed short-time-increment storm sequences. Journal of Hydrology 16:307–315.

Williams, G. P. 1984. Paleohydrologic equations for rivers. In: J. E. Costa and P. J. Fleishcher (ed.), Developments and Applications of Geomorphology, pp. 353–367. Springer-Verlag, Berlin.

Wilson, C. B., J. B. Valdes, and I. Rodriguez-Iturbe. 1979. On the influence of the spatial distribution of rainfall on storm runoff. Water Resources Research 15(2):321–322.

Wiltshire, S. E. 1986a. Identification of homogeneous regions for flood frequency analysis. Journal of Hydrology 84:287–302.

Wiltshire, S. E. 1986b. Regional flood frequency analysis I: Homogeneity statistics. Journal of Hydrological Sciences.

Wiltshire, S. E. 1986c. Regional flood frequency analysis II: Multivariate classification of drainage basins in Britain. Journal of Hydrological Sciences.

World Meteorological Organization. 1982 (rev. 1986). Manual for Estimation of Probable Maximum Precipitation. Operational Hydrology Report No. 1, WMO No. 232. Geneva, Switzerland. 269 pp.

Yankee Atomic Electric Company. 1984. Probability of extreme rainfalls and the effect of the Harriman Dam, Framingham, Massachusetts. 16 pp.

Zawadzki, I. I. 1973. Statistical properties of precipitation patterns. Journal Applied Meteorology 12:459–472.

Zhang, Y. 1982. Plotting positions of annual flood extremes considering extraordinary values. Water Resources Research 19(4):859–864.

Biographical Sketches of Committee Members

JARED L. COHON (*Chairman*) received his B.S. in Civil Engineering from the University of Pennsylvania in 1969 and his S.M. and Ph.D. in Civil Engineering (water resources) from MIT in 1972 and 1973, respectively. Currently he is Vice-Provost for Research and Professor of Geography and Environmental Engineering at The Johns Hopkins University. Previously, he was Associate Dean of Engineering, Associate Professor, and Assistant Professor at Hopkins and Legislative Assistant to Senator Daniel P. Moynihan. His principal interests are water resources planning and management and multiobjective optimization in water resources systems, topics on which he has published extensively.

VICTOR R. BAKER received his B.S. in Geology at the Rensselaer Polytechnic Institute in 1967 and his Ph.D. in Geology at the University of Colorado in 1971. Currently he is Professor of Geosciences and Professor of Planetary Sciences at the Lunar and Planetary Laboratory, The University of Arizona. Previously, he was a Hydrologist/Geophysicist for the U.S. Geological Survey; City Geologist of Boulder, Colo.; and a Research Scientist for the Bureau of Economic Geology, The University of Texas. Dr. Baker is widely known for his work relevant to the application of paleohydrologic methods in the analysis of floods.

DUANE C. BOES received his B.A. at St. Ambrose College, Davenport, Iowa, in 1956 and his M.S. and Ph.D. at Purdue University in 1958 and 1963, respectively. Currently he is Chairman of the Department of Statistics at the Colorado State University. He has nearly 30 years experience in the field of statistics. His principal interests are stochastic modeling and time series analysis of geophysical phenomena, statistical inference, and reservoir and storage theory.

C. ALLIN CORNELL received his A.B. in Architecture (1960) and M.S. (1961) and Ph.D. (1964) in Civil Engineering (Structures), all from Stanford University. Currently he is Professor in the Department of Civil Engineering at Stanford. Previously, he was on the faculty of MIT for 20 years. Dr. Cornell has been on the forefront of research in probabilistically based engineering and design for structural safety. For example, he developed the first formal consistent method for calculating probability of occurrence of different levels of earthquake ground motion. In addition to his academic and research achievements, Dr. Cornell consults widely on such problems as seismic considerations for nuclear plants, wind loads, and wind-induced wave risk. Dr. Cornell has been a member of the National Academy of Engineering since 1981.

NORMAN H. CRAWFORD received his B.S. in Civil Engineering in 1958 from the University of Alberta, Canada, and his M.S. and Ph.D. in Hydraulic Engineering from Stanford University in 1959 and 1962, respectively. Currently he is President of Hydrocomp. His professional specialties are in the following areas: development of computer simulation methods for streamflow, sediment transport, and aquatic ecology; applications of continuous hydrologic simulation to water resource management including hydropower project design, operational forecasting and optimization; applications of continuous simulation of hydrology and water quality in regional water resource and environmental planning; and economic analysis of watershed development alternatives. Previously, he was Assistant Professor of Hydrologic Engineering at Stanford University.

MICHAEL D. HUDLOW received his B.S. in General Engineering in 1963 at the Texas A&I University and his M.S. and Ph.D. in Hydrometeorology at the Texas A&M University in 1966 and 1967, respectively. Currently, he is Director, Office of Hydrology, National Weather Service. Formerly, he was Chief of the Hydrologic Research Laboratory, Office of Hydrology at the National Weather Service. He has served in various hydrometeorological research and supervisory positions in NOAA since 1969 and prior to that in several atmospheric components of the U.S. Army. He has been a member of the American Meteorological Society (AMS) since 1965 and has been active in radar meteorology, tropical meteorology, and hydrometeorological activities of the AMS.

WILLIAM KIRBY holds B.C.E. (1963), M.S. (1966), and Ph.D. (1968) degrees, all from Cornell University, in Sanitary Engineering, Hydraulics, and Applied Probability. He has worked as a research hydrologist for the U.S. Geological Survey since 1967; currently he is a hydrologist in the Office of Surface Water, where he develops and maintains procedures and computer programs for indirect discharge determinations and other hydraulic computations and develops procedures for calculating probability laws of hydrologic storage models for floods and droughts. He has had considerable experience in watershed modeling and flood-frequency analysis.

DONALD W. NEWTON received his B.S. in 1949 from Antioch College and his M.S. in 1951 from the University of Colorado in Civil Engineering. He currently is supervisor of Hydrology Section of the Flood Protection Branch of the Tennessee Valley Authority (since 1962). His expertise is in flood hydrology, hydrologic design of water resources structures, and water resources planning and management. He currently chairs an American Society of Civil Engineers Task Committee on Spillway Design Floods.

KENNETH W. POTTER received his B.S. in Geology from Louisiana State University in 1968 and his Ph.D. in Geography and Environmental Engineering from The Johns Hopkins University in 1976. Currently he is Professor at the University of Wisconsin-Madison, where he teaches courses in hydrology and water resources. His research largely involves flood hydrology and hydrologic design, topics on which he has published widely. Dr. Potter is also a member of the WSTB Committee on Water Resources Research.

JAMES R. WALLIS received his B.S. in forestry from the University of New Brunswick in 1950, M.S. from Oregon State University of 1954, and Ph.D. in soil morphology from the University of California, Berkeley in 1965. Currently he is a Research Staff Member at the IBM Thomas J. Watson Research Center, where he has been since 1967. Previously, he held positions in hydrology and forestry with the U.S. Forest Service, Montana State University, and elsewhere. His principal interests are in mathematical models applied to hydrology, soils, forestry, and land management. Dr. Wallis has lectured at many different universities and has addressed many issues relevant to estimates of extreme floods.

SIDNEY J. YAKOWITZ received his B.S. in Electrical Engineering from Stanford University in 1960 and his M.S. and Ph.D. in Electrical Engineering from Arizona State University in 1966 and 1967, respectively. Currently he is Professor with the Department of Systems and Industrial Engineering at the University of Arizona. Previously, he was a Design Engineer at Lawrence Radiation Laboratory at the University of California, Berkeley.

STEPHEN J. BURGES (*ex-officio*) received his B.Sc. in Physics and Mathematics and B.E. in Civil Engineering at the University of Newcastle, Australia, in 1967. He received an M.S. (1968) and Ph.D. (1970) in Civil Engineering from Stanford University. He has been a member of the faculty at the University of Washington since 1970 and currently is a professor of civil engineering. Dr. Burges is also a member of the Water Science and Technology Board.

LEO M. EISEL (*ex-officio*) is a hydrologist who received his Ph.D. in Engineering from Harvard University in 1970. Currently Dr. Eisel is Vice President of Wright Water Engineers in Denver, Colo. Previously he was Director of the U.S. Water Resources Council, Staff Scientist with the Environmental Defense Fund in New York, and Director of the Illinois Division of Water Resources, positions that involved responsibility for evaluation, development, and management of water resources for flood control, water quality, water supply, power navigation, recreation, and fish and wildlife conservation. Dr. Eisel was a member of the Water Science and Technology Board through June 1987.

Index

A

Agricultural Research Service (ARS), 93
Air temperature, 104
Alexander's storm transposition method, 70–71
Antecdent conditions
 specification of, 56, 58
 stochastic model of, 74
ARS (Agricultural Research Service), 93
At-site analysis, 8
Attraction, domain of, 19

B

Basin characteristics, 38
Bayesian analysis, 53
Bias of estimators, 16

C

Calibration-curve methods, 89
Calibration process, 58
Central limit theorem, 27
Channel geometry, 108
Climatic change, 114

Comprehensive statistical model, 122
Confidence interval estimator, 17
Correlation coefficient, 24–25
Covariance, 24
Current-meter measurements, 90
Cv (coefficient of variation), 41, 44

D

D–A–D (depth–area–duration) matrix, 96–97, 98
Daily value files, 86–87
Data characteristics and availability, 80–117
Data collection platforms (DCPs), 94
Data error, 75–77
DCPs (data collection platforms), 94
Depth–area–duration (D–A–D) matrix, 96–97, 99
Depth–duration–frequency relationship method, 63–64
Discharge models, *see* Runoff models

E

Earthquakes, 117
Efficiency of estimators, 17, 20
Error, kinds of, 75-77

Estimators, 16
 efficiency of, 17, 20
 moment, 22-23
 procedures for finding, 17
 unbiased, 36-37
 see also Floods; Probabilities of extreme floods
Evapotranspiration, 104
Evapotranspiration values, 72
Exceedance probabilities, 63
Extremal types theorem, 19
Extreme events, principle of focus on, 7
Extreme floods, probabilities of, *see* Probabilities of extreme floods
Extreme-value theory, 13, 18-19

F

Fisher information matrix, 17-18
Floods
 characteristics of, 111-117
 probability estimation, *see* Probabilities of extreme floods
 probable maximum (PMF), 2, 5
 statistical techniques and analysis, 12-54, 118-119
Floodset descriptions, 40-52

G

Gages, rain, 26, 96
Generalized extreme value distribution, *see* GEV *entries*
Generic parameter, 15
Geographic regional analysis, 9
Geomorphic instantaneous unit hydrograph, 73
Geostationary Operational Environmental Satellite (GOES), 94
GEV (generalized extreme value distributions), 8, 19, 20
GEV-1 growth curve estimates, 51, 52
GEV-1 index flood model, 40, 44
GEV-1 quantile estimates, 45, 47, 48, 50, 52
GEV-2 index flood model, 40, 44
GEV-2 quantile estimates, 46, 47, 49

GOES (Geostationary Operational Environmental Satellite), 94
Goodness, measures of, 16

H

Historical data, 32-34, 53
 general characteristics of, 105-111
 storm catalog, 10, 96-100
 treatment of, 115-117
Hourly Precipitation Data, 93
Hydrology Subcommittee, 2
Hydrometeorologic information, 104-105

I

IDFs (intensity-duration-frequency curves), 7
Index flood methods, 38-40
Infiltration modeling, 77
Infimum, 14
Intensity-duration-frequency curves (IDFs), 7
Interstation correlations, 113
Interval estimation, 12
Isohyetal maps, 97, 99

J

Joint distribution, 14

L

Large floods, *see* Probabilities of extreme floods
Likelihood function, 32-34
Log-Pearson Type III distribution, 8

M

Manually digitized radar (MDR), 101-102
Maximum flood, probable (PMF), 2, 5
Maximum likelihood estimation (MLE), 17-18, 27-31
Maximum precipitation, probable (PMP), 5, 64

INDEX 139

MDR (manually digitized radar), 101–102
Mean square error (MSE), 16
Meteorologic homogeneity, 68–69
Meteorologic inputs, 78–79
  development of, 58–73
MLE (maximum likelihood estimation), 17–18, 27–31
Models and modeling, defined, 13, 14; *see also specific types of models*
Moment estimators, 22–23
Monte Carlo experiments, 40–41
MSE (mean square error), 16
Multidimensional space-time rainfall models, 60–62
Multisite regional anlysis, 34–52
Multivariate log normal, 35–39

N

National Center for Atmospheric Research (NCAR), 102
National Climatic Data Center (NCDC), 92–93
National Environmental Data Referral Service (NEDRES), 92
National Environmental Satellite Data and Information Service (NESDIS), 92, 104
National Oceanic and Atmospheric Administration (NOAA), 92
National Water Data Exchange (NAWDEX), 84–85, 88, 92
National Weather Service (NWS), 92
NAWDEX (National Water Data Exchange), 84–85, 88, 92
NCAR (National Center for Atmospheric Research), 102
NCDC (National Climatic Data Center) , 92–93
NEDRES (National Environmental Data Referral Service), 92
NESDIS (National Environmental Satellite Data and Information Service), 92, 104
Next Generation Weather Radar (NEXRAD), 102
NOAA (National Oceanic and Atmospheric Administration), 92
Nonparametric procedures, 31

Nonstationary records, 113–114
Nonsystematic records, 82–84
NWS (National Weather Service), 92

P

Paleoflood data, *see* Historical data
Paleogeomorphic-based flow data, 106–111
Paleohydrology, 105
Paleostage-based flow data, 107–111
Parameter error, 75–77
Paremter space, 15
Parametric modeling, 54
Pareto-type distribution, 31
Partial–duration series, 85
PMF (probable maximum flood), 2, 5
PMP (probable maximum precipitation), 5, 64
Point estimation, 12
Precipitaton
  input, errors in, 75–76
  major events, distribution of, 66, 67
  probable maximum (PMP), 5, 64
  see also Rainfall *entries*; Snow *entries*
Probabilities of extreme floods
  approaches to estimation of, 4–6
  errors in estimation of, 77
  estimation of, 12–13, 73–74, 79
  principles for improving estimation of, 6–7
Probability weighted moments (PWMs), 20–22
  regional, 39
Probable maximum flood (PMF), 2, 5
Probable maximum precipitation (PMP), 5, 64
PWMs, *see* Probability weighted moments

R

RADAP II (Radar Data Processor II), 101–103
Radar rainfall data, 101–103
Radiocarbon dating, 109
Rain gages, 26, 96
Rainfall data, 91–105

accuracy and interpretation of, 94–95
actual, direct use of, 59–60
availability of, 92–94
developmental issues in, 120–122
inputs, 58–65
radar, 101–103
systematic versus nonsystematic records of, 81–84
Rainfall models
multidimensional space-time, 60–62
stochastic, 60–62
see also Runoff models
Rainfall-runoff models, *see* Runoff models
Regime-based paleoflow estimates (RBPE), 106
Regional analysis, multisite, 34–52
Regional quantile functions, 39
Regionalization, 6, 13
by regression methods, 38
runoff models and, 56
studies, 37–39
techniques, 118–119
Research needs, 118-122
Robustness, 13
quantified, 20
Runoff models, 78; *see also* Rainfall models
continuous, 56
defined, 57–58
developmental issues and research needs, 119–120
errors in, 75–77
methods for, 55–79
recommended, 9-11
regionalization and, 56

S

Sampling errors, 25–26
Satellite data, 103–104
SCS (Soil Conservation Service), 101
Shear stress, velocity, and stream power (SS-V-SP) studies, 106–107
Significant storm, defined, 65–68
Single-site frequency estimation, 26–34
Skew coefficient, 23
Skew-kurtosis relationship, 23, 25

Slackwater deposition and paleostage indicators (SWD-PSI), 107, 109–110
Slope-area analysis, 107
Snow data, 100-101
Snowpack melt, 11
Soil Conservation Service (SCS), 101
Space-time rainfall models, multidimensional, 60–62
SS–V–SP (shear stress, velocity, and stream power) studies, 106–107
Standard error of estimate, 17
Station-header file, 85
Statistical techniques
flood-based, 12–54
model, comprehensive, 122
recommended, 8–9
sampling conditions, 115–117
Step-backwater analysis, 107
Stochastic models, 14, 60–62, 74
Storm Rainfall in the United States, 10, 96–100
Storms
historical catalog, 10, 96–100
interior, 62
significant, defined, 65–68
synthetic, 10, 62–65
transposition methods, 64–72
Stratigraphy, 109
Streamflow, 13–14
Streamflow-basin characteristics file, 85, 87
Streamflow data, 2, 84–91
accuracy of, 90–91
availability of, 84–88
collection and computation principles, 89
developmental issues in, 120–122
paleogeomorphic-based, 106–111
paleostage-based, 107–111
supplementary data files, 87–88
systematic versus nonsystematic records of, 81–84
time-series data files, 85–87
types of, 85–88
Substitution of space for time, principle of, 6
Supremum, 14
SWD–PSI (slackwater deposition and paleostage indicators), 107, 109–110

Synthetic storms, 10, 62–65
Systematic records, 82, 84

T

Tail behavior results, 29–31
Tail probability estimation, 12, 16
Tail quantile estimation, 12–13, 16
Time scale, 14

U

Unbiased estimators, 36–37
Uncertainty analysis, 11, 74–77, 79
Unit-values file, 86
U.S. Army Corps of Engineers, 94, 96
U.S. Bureau of Reclamation, 100
U.S. Geological Survey (USGS), 84, 88
U.S. Nuclear Regulatory Commission, 1, 2

V

Variance, asymptotic, 28–29
Variation, coefficient of (Cv), 41, 44

W

Wakeby distributions, 8
WAK/R index flood model, 39–40, 44
WAK/R quantile estimates, 46
Water Data Storage and Retrieval System (WATSTORE), 84–87
Wind, errors caused by, 95

Y

Yankee Atomic Electric Company storm transposition method, 71–72